on the
SHOULDERS
of GIANTS

on the SHOULDERS of GIANTS

A COURSE IN SINGLE VARIABLE CALCULUS

GH SMITH + GJ McLELLAND

A UNSW Press book

Published by
University of New South Wales Press Ltd
University of New South Wales
UNSW Sydney NSW 2052
AUSTRALIA
www.unswpress.com.au

© GH Smith and GJ McLelland
First published 2003

This book is copyright. Apart from any fair dealing for the purpose of private study, research, criticism or review, as permitted under the Copyright Act, no part may be reproduced by any process without written permission. Inquiries should be addressed to the publisher.

National Library of Australia
Cataloguing-in-Publication entry:

> Smith, Geoff, 1953– .
> On the shoulders of giants: a course in single variable calculus.
>
> [Rev. ed.].
> Includes index.
> ISBN 0 86840 717 8.
> 1. Calculus of variations. 2. Engineering mathematics.
> 3. Science — Mathematics. I. McLelland, G.J. II. Title.

515.64

Printer BPA Print Group, Melbourne
Illustrations pages 2, 5, 8 and 195 Anita Howard
Cover design Di Quick

CONTENTS

Preface   v

1 Terror, tragedy and bad vibrations   1
   1.1 Introduction 1
   1.2 The *Tower of Terror* 1
   1.3 Into thin air 4
   1.4 Music and the bridge 7
   1.5 Discussion 8
   1.6 Rules of calculation 9

2 Functions   11
   2.1 Rules of calculation 11
   2.2 Intervals on the real line 15
   2.3 Graphs of functions 17
   2.4 Examples of functions 20

3 Continuity and smoothness   27
   3.1 Smooth functions 27
   3.2 Continuity 30

4 Differentiation   41
   4.1 The derivative 41
   4.2 Rules for differentiation 48
   4.3 Velocity, acceleration and rates of change 53

5 Falling bodies   57
   5.1 The *Tower of Terror* 57
   5.2 Solving differential equations 62
   5.3 General remarks 65
   5.4 Increasing and decreasing functions 68
   5.5 Extreme values 70

6 Series and the exponential function   75
   6.1 The air pressure problem 75
   6.2 Infinite series 81
   6.3 Convergence of series 84
   6.4 Radius of convergence 90

6.5	Differentiation of power series	93
6.6	The chain rule	96
6.7	Properties of the exponential function	99
6.8	Solution of the air pressure problem	102

7 Trigonometric functions   109

7.1	Vibrating strings and cables	109
7.2	Trigonometric functions	111
7.3	More on the sine and cosine functions	114
7.4	Triangles, circles and the number π	119
7.5	Exact values of the sine and cosine functions	122
7.6	Other trigonometric functions	125

8 Oscillation problems   127

8.1	Second order linear differential equations	127
8.2	Complex numbers	134
8.3	Complex series	140
8.4	Complex roots of the auxiliary equation	143
8.5	Simple harmonic motion and damping	145
8.6	Forced oscillations	153

9 Integration   167

9.1	Another problem on the *Tower of Terror*	167
9.2	More on air pressure	168
9.3	Integrals and primitive functions	170
9.4	Areas under curves	171
9.5	Area functions	174
9.6	Integration	176
9.7	Evaluation of integrals	182
9.8	The fundamental theorem of the calculus	187
9.9	The logarithm function	188

10 Inverse functions   197

10.1	The existence of inverses	200
10.2	Calculating function values for inverses	205
10.3	The oscillation problem again	214
10.4	Inverse trigonometric functions	218
10.5	Other inverse trigonometric functions	221

11 Hyperbolic functions   225

11.1	Hyperbolic functions	225
11.2	Properties of the hyperbolic functions	227
11.3	Inverse hyperbolic functions	230

12 Methods of integration — 235
- 12.1 Introduction — 235
- 12.2 Calculation of definite integrals — 237
- 12.3 Integration by substitution — 239
- 12.4 Integration by parts — 241
- 12.5 The method of partial fractions — 243
- 12.6 Integrals with a quadratic denominator — 247
- 12.7 Concluding remarks — 249

13 A nonlinear differential equation — 251
- 13.1 The energy equation — 252
- 13.2 Conclusion — 259

Answers — 261

Index — 281

PREFACE

> If I have seen further it is by standing on the shoulders of Giants.
>
> *Sir Isaac Newton, 1675.*

This book presents an innovative treatment of single variable calculus designed as an introductory mathematics textbook for engineering and science students. The subject material is developed by modelling physical problems, some of which would normally be encountered by students as experiments in a first year physics course. The solutions of these problems provide a means of introducing mathematical concepts as they are needed. The book presents all of the material from a traditional first year calculus course, but it will appear for different purposes and in a different order from standard treatments.

The rationale of the book is that the mathematics should be introduced in a context tailored to the needs of the audience. Each mathematical concept is introduced only when it is needed to solve a particular practical problem, so at all stages, the student should be able to connect the mathematical concept with a particular physical idea or problem. For various reasons, notions such as *relevance* or *just in time mathematics* are common catchcries. We have responded to these in a way which maintains the professional integrity of the courses we teach.

The book begins with a collection of problems. A discussion of these problems leads to the idea of a function, which in the first instance will be regarded as a rule for numerical calculation. In some cases, real or hypothetical results will be presented, from which the function can be deduced. Part of the purpose of the book is to assist students in learning how to define the rules for calculating functions and to understand why such rules are needed. The most common way of expressing a rule is by means of an algebraic formula and this is the way in which most students first encounter functions. Unfortunately, many of them are unable to progress beyond the *functions as formulas* concept. Our stance in this book is that functions are rules for numerical calculation and so must be presented in a form which allows function values to be calculated in decimal form to an arbitrary degree of accuracy. For this reason, trigonometric functions first appear as power series solutions to differential equations, rather than through the common definitions in terms of triangles. The latter definitions may be intuitively simpler, but they are of little use in calculating function values or preparing the student for later work. We begin with simple functions defined by algebraic formulas and move on to functions defined by power series and integrals. As we progress through the book, different physical problems give rise to various functions and if the calculation of function values requires the numerical evaluation of an integral, then this simply has to be accepted as an inconvenient but unavoidable property of the problem. We would like students to appreciate the fact that some problems, such as the nonlinear pendulum, require sophisticated mathematical methods for their analysis and difficult mathematics is unavoidable if we wish to solve the problem. It is not introduced simply to provide an intellectual challenge or to filter out the weaker students.

Our attitude to proofs and rigour is that we believe that all results should be correctly stated, but not all of them need formal proof. Most of all, we do not believe that students should be presented with handwaving arguments masquerading as proofs. If we feel that a proof is accessible and that there is something useful to be learned from the proof, then we provide it. Otherwise, we state the result and move on. Students are quite capable of using the results on term-by-term differentiation of a power series for instance, even if they have not seen the proof. However, we think that it is important to emphasise that a power series can be differentiated in this way only within the interior of its interval of convergence. By this means we can take the applications in this book beyond the artificial examples often seen in standard texts.

We discuss continuity and differentiation in terms of convergence of sequences. We think that this is intuitively more accessible than the usual approach of considering limits of functions. If limits are treated with the full rigour of the ε-δ approach, then they are too difficult for the average beginning student, while a non-rigorous treatment simply leads to confusion.

The remainder of this preface summarises the content of this book. Our list of physical problems includes the vertical motion of a projectile, the variation of atmospheric pressure with height, the motion of a body in simple harmonic motion, underdamped and overdamped oscillations, forced damped oscillations and the nonlinear pendulum. In each case the solution is a function which relates two variables. An appeal to the student's physical intuition suggests that the graphs of these functions should have certain properties. Closer analysis of these intuitive ideas leads to the concepts of continuity and differentiability. Modelling the problems leads to differential equations for the desired functions and in solving these equations we discuss power series, radius of convergence and term-by-term differentiation. In discussing oscillation we have to consider the case where the auxiliary equation may have non-real roots and it is at this point that we introduce complex numbers. Not all differential equations are amenable to a solution by power series and integration is developed as a method to deal with these cases. Along the way it is necessary to use the chain rule, to define functions by integrals and to define inverse functions. Methods of integration are introduced as a practical alternative to numerical methods for evaluating integrals if a primitive function can be found. We also need to know whether a function defined by an integral is new or whether it is a known elementary function in another form. We do not go very deeply into this topic. With the advent of symbolic manipulation packages such as *Mathematica*, there seems to be little need for science and engineering students to spend time evaluating anything but the simplest of integrals by hand. The book concludes with a capstone discussion of the nonlinear equation of motion of the simple pendulum. Our purpose here is to demonstrate the fact that there are physical problems which absolutely need the mathematics developed in this book. Various *ad hoc* procedures which might have sufficed for some of the earlier problems are no longer useful. The use of *Mathematica* makes plotting of elliptic functions and finding their values no more difficult than is the case with any of the common functions.

We would like to thank Tim Langtry for help with LaTeX. Tim Langtry and Graeme Cohen read the text of the preliminary edition of this book with meticulous attention and made numerous suggestions, comments and corrections. Other useful suggestions, contributions and corrections came from Mary Coupland and Leigh Wood.

CHAPTER 1
TERROR, TRAGEDY AND BAD VIBRATIONS

1.1 INTRODUCTION

Mathematics is almost universally regarded as a useful subject, but the truth of the matter is that mathematics beyond the middle levels of high school is almost never used by the ordinary person. Certainly, simple arithmetic is needed to live a normal life in developed societies, but when would we ever use algebra or calculus? In mathematics, as in many other areas of knowledge, we can often get by with a less than complete understanding of the processes. People do not have to understand how a car, a computer or a mobile phone works in order to make use of them. However, some people do have to understand the underlying principles of such devices in order to invent them in the first place, to improve their design or to repair them. Most people do not need to know how to organise the Olympic Games, schedule baggage handlers for an international airline or analyse traffic flow in a communications network, but once again, someone must design the systems which enable these activities to be carried out. The complex technical, social and financial systems used by our modern society all rely on mathematics to a greater or lesser extent and we need skilled people such as engineers, scientists and economists to manage them. Mathematics is widely used, but this use is not always evident. Part of the purpose of this book is to demonstrate the way that mathematics pervades many aspects of our lives. To do this, we shall make use of three easily understood and obviously relevant problems. By exploring each of these in increasing detail we will find it necessary to introduce a large number of mathematical techniques in order to obtain solutions to the problems. As we become more familiar with the mathematics we develop, we shall find that it is not limited to the original problems, but is applicable to many other situations.

In this chapter, we will consider three problems: an amusement park ride known as the *Tower of Terror*, the disastrous consequences that occurred when an aircraft cargo door flew open in mid-air and an unexpected noise pollution problem on a new bridge. These problems will be used as the basis for introducing new mathematical ideas and in later chapters we will apply these ideas to the solution of other problems.

1.2 THE *TOWER OF TERROR*

Sixteen people are strapped into seats in a six tonne carriage at rest on a horizontal metal track. The power is switched on and in six seconds they are travelling at 160 km/hr. The carriage traverses a short curved track and then hurtles vertically upwards to reach the height of a 38 storey building. It comes momentarily to rest and then free falls for about five seconds to again reach a speed of almost

Figure 1.1: The *Tower of Terror*

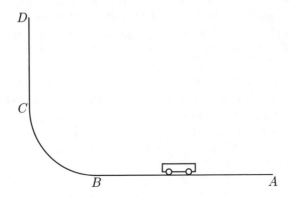

Figure 1.2: The *Tower of Terror* (Schematic)

160 km/hr. It hurtles back around the curve to the horizontal track where powerful brakes bring it to rest back at the start. The whole event takes about 25 seconds (Figure 1.1).

This hair-raising journey takes place every few minutes at Dreamworld, a large amusement park on the Gold Coast in Queensland, Australia. Parks like this have become common around the world with the best known being Disneyland in the United States. One of the main features of the parks are the rides which are offered and as a result of competition between parks and the need to continually change the rides, they have become larger, faster and more exciting. The ride just described is aptly named the *Tower of Terror*.

These trends have resulted in the development of a specialised industry to develop and test the rides which the parks offer. There are two aspects to this. First the construction must ensure that the equipment will not collapse under the strains imposed on it. Such failure, with the resulting shower of fast-moving debris over the park, would be disastrous. Second, and equally important, is the need to ensure that patrons will be able to physically withstand the forces to which they will be subjected. In fact, many rides have restrictions on who can take the ride and there are often warning notices about the danger of taking the ride for people with various medical problems.

Let's look at some aspects of the ride in the *Tower of Terror* illustrated in the schematic diagram in Figure 1.2. The carriage is accelerated along a horizontal track from the starting point A. When it reaches B after about six seconds, it is travelling at 160 km/hr and it then travels around a curved portion of track until its motion has become vertical by the point C. From C the speed decreases under the influence of gravity until it comes momentarily to rest at D, 115 metres above the ground or the height of a 38 storey building. The motion is then reversed as the carriage free falls back to C. During this portion of the ride, the riders experience the sensation of weightlessness for five or six seconds. The carriage then goes round the curved section of the track to reach the horizontal portion of the track, the brakes are applied at B and the carriage comes to a stop at A.

The most important feature of the ride is perhaps the time taken for the carriage to travel from D back to C. This is the time during which the riders experience weightlessness during free fall. If the time is too short then the ride would be pointless. The longer the time however, the higher the tower must be, with the consequent increase in cost and difficulty of construction. The time depends on the speed at which the carriage is travelling when it reaches C on the outward journey and the higher this speed the longer the horizontal portion of the track must be and the more power is required to accelerate the carriage on each ride. The design of the ride is thus a compromise between the time taken for the descent, the cost of construction and the power consumed on each ride.

The first task is to find the relation between the speed at C, the height of D and the time taken to travel from D to C. This is a modern version of the problem of the motion of falling bodies, a problem which has been discussed for about 2,500 years.

The development of the law of falling bodies began in Greece about 300 BC. At that time Greece was the intellectual centre of the western world and there were already two hundred years of scientific and philosophical thought to build upon. From observation of everyday motions, the Greek philosopher Aristotle put forward a collection of results about the motion of falling bodies as part of a very large system of ideas that was intended to explain the whole of reality as it was then known. Other Greek thinkers were also producing such ambitious systems of ideas, but Aristotle was the only one to place much importance on the analysis of motion as we would now understand the word.

Almost all of Aristotle's methods for analysing motion have turned out to be wrong, but he was nevertheless the first to introduce the idea that motion could be analysed in numerical terms. Aristotle's ideas about motion went almost unchallenged for many centuries and it was not until the 14th century that a new approach to many of the problems of physics began to emerge. Perhaps the first real physicist in modern terms was Galileo, who in 1638 published his *Dialogues Concerning the Two New Sciences* in which he presented his ideas on the principles of mechanics. He was the first person to give an accurate explanation of the motion of falling bodies in more or less modern terms. With nobody to show him how to solve the problem, it required great insight on his part to do this. But once Galileo had done the hard work, everybody could see that the problem was an easy one to solve and it is now a routine secondary school exercise. We shall derive the law from a hypothetical set of experimental results to illustrate the way in which mathematical methods develop.

1.3 INTO THIN AIR

At 1.33 a.m. on 24 February 1989 flight UAL 811 left Honolulu International Airport bound for Auckland and Sydney with 337 passengers and 18 crew on board. About half an hour later, when the aircraft was over the ocean 138 km south of the airport and climbing through 6700 m, the forward cargo door opened without warning, and was torn off, along with 7 square metres of the fuselage. As a result of this event, there was an outrush of air from the cabin with such force that nine passengers were sucked out and never seen again. The two forward toilet compartments were displaced by 30 cm. Two of the engines and parts of the starboard wing were damaged by objects emerging from the aircraft and the engines had to be shut down. The aircraft turned back to Honolulu and, with considerable difficulty, landed at 2.34 a.m. Six tyres blew out during the landing and the brakes seized. All ten emergency slides were used to evacuate the passengers and crew and this was achieved with only a few minor injuries.

As with all aircraft accidents, extensive enquiries were conducted to find the cause. A coast guard search under the flight path located 57 pieces of material from the aircraft, but no bodies were found. The cargo door was located and recovered in two separate pieces by a United States Navy submarine in October 1990. After inspecting the door and considering all other evidence, the US National Transportation Safety Board concluded that a faulty switch in the door control system had caused it to open. The Board made recommendations about procedures which would prevent such accidents in future and stated that proper corrective action after a similar cargo door incident in 1987 could have prevented the tragedy.

The event which triggered the accident was the opening of the cargo door, but the physical cause of the subsequent events was the explosive venting of air from the aircraft. The strong current of air

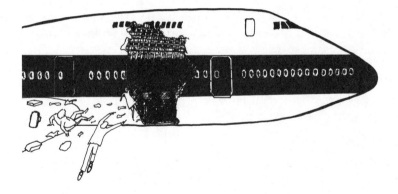

Figure 1.3: Flight UAL811

was apparent to all on board. After the initial outrush of air, the situation in the aircraft stabilised, but passengers found it difficult to breathe. A first attempt to explain this event might be that the speed of the aircraft through the still air outside caused the air inside to be sucked out. There are several reasons why this is not convincing. Firstly, the phenomenon does not occur at low altitudes. If a window is opened in a fast moving car or a low flying light aircraft, the air is not sucked out. Secondly, the same breathing difficulties are experienced on high mountains when no motion at all is taking place. It appears that the atmosphere becomes thinner in some way as height increases, and that, as a result, it is difficult to inhale sufficient air by normal breathing. In addition, if air at normal sea level pressure is brought in contact with the thin upper level air, as occurred in the accident with flight 811, there will be a flow of air into the region where the air is thin.

The physical mechanism which is at the heart of the events described above is also involved in a much less dramatic phenomenon and it was in this other situation that the explanation of the mechanism first emerged historically.

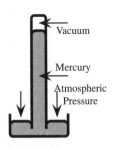

Figure 1.4: Barometer

In 1643 Evangelista Torricelli, a friend and follower of Galileo, constructed the first modern barometer. This is shown schematically in Figure 1.4. Torricelli took a long glass tube and filled it with mercury. He closed the open end with a finger and then inverted the tube with the open end in a vessel containing mercury. When the finger was released, the mercury in the tube always dropped to a level of about 76 cm above the mercury surface in the open vessel. The density of mercury is 13.6 gm/cm^3 and so the weight of a column of mercury of unit area and 76 cm high is 1030 gm. This weight is almost identical to the weight of a column of water of unit area and height 10.4 m, given that the density of water is 1 gm/cm^3. In fact, water barometers had been constructed a few years before Torricelli and it had been found the maximum height of the column of water was 10.4 m.

The simplest way to describe these experiments is in terms of the pressure exerted by the column of fluid, whether air, water or mercury. As shown in Figure 1.4, the weight of the liquid in the tube is exactly balanced by the weight of the atmosphere pushing down on the liquid surface.

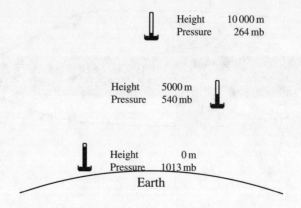

Figure 1.5: Variation of pressure with height

It was soon found that atmospheric pressure is not constant even at sea level and that the small variations in pressure are related to changes in the weather. It was also found that the pressure decreases with height above sea level and this is to be expected since the mass of the column of air decreases with height (Figure 1.5).

We can now give a partial explanation of the events of Flight 811. As aircraft cabins are pressurised, the pressure inside the cabin would have been approximately that of normal ground level pressure. The external pressure would have been less than half this value. When the cargo door burst open, the internal pressure forced air in the aircraft out of the opening until the internal and external pressures were equal, at which time the situation stabilised. The difficulty in breathing would have been due to the reduced pressure, since we need this pressure to force air into our lungs.

It is essential to have some model for the variation of pressure with height because of the needs of weather forecasting, aircraft design, mountaineering and so on, but the variation of atmospheric pressure with height does not follow a simple rule. As with the falling body problem, a set of experimental results will be used to obtain at least an approximate form for the required law. These results will be the average value for the pressure at various heights in the atmosphere. Obtaining the law of pressure variation from this set of experimental results will be more difficult than in the case of a falling body.

EXERCISES 1.3

1. There are many common devices which utilise fluid pressure. Examples include dentists' chairs, car lifts and hydraulic brakes. What other examples can you think of?

2. It is known from physical principles that the pressure exerted at a depth d in an incompressible fluid (such as water) is given by $P = \rho g d \, \text{N/m}^2$, where ρ is the density of the fluid and g is the acceleration due to gravity—approximately 9.81 m/s^2. A swimming pool is 8 m long, 5 m wide and 2 m deep. What force is exerted on the bottom of the pool by the weight of the water? (You may take $\rho = 1000 \, \text{kg/m}^3$.) Suppose the pool were filled with seawater ($\rho = 1030 \, \text{kg/m}^3$). What force is now exerted on the bottom of the pool by the weight of the water?

3. The column of mercury in a barometer is 75 cm high. Compute the air pressure in kg/m^2.

4. A pump is a device which occurs in many situations—pumping fuel in automobiles, pumping water from a tank or borehole or pumping gas in air conditioners or refrigerators. A simple type of water pump is shown in the diagram on the left. A cylinder containing a piston is lowered into a tank. The cylinder has a valve at its lower end and there are valves on the piston. When the piston is moving down, the valve in the cylinder closes and the valves in the piston open. When the piston moves up, the valve in the cylinder opens and the valves in the piston close. Based on the discussion on air pressure given in the text, explain how such a system can be used to pump water from the tank. Explain also why the maximum height to which water can be raised with such a pump is about 10.4 m.

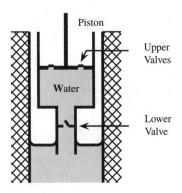

1.4 MUSIC AND THE BRIDGE

Almost everyone enjoys a quiet night at home, but in the modern world there is less and less opportunity for this simple pleasure. Whether it is aircraft noise, loud parties, traffic din or sporting events, there are many forms of noise pollution which cause annoyance or disruption. New forms of noise pollution are continually arising and some of these are quite unexpected.

Sydney contains a large number of bridges, the best known of these being the Sydney Harbour Bridge. The newest bridge is the A$170m Anzac Bridge, originally known as the Glebe Island Bridge (Figure 1.6). The main span of this concrete bridge is 345 m long and 32.2 m wide. The deck is supported by two planes of stay cables attached to two 120 m high reinforced concrete towers. It is the cables which created an unexpected problem. As originally designed, they were enclosed in polyurethane coverings. When the wind was at a certain speed from the south-east, the cables began to vibrate and bang against the coverings. The resulting noise could be heard several kilometres away from the bridge, much to the annoyance of local residents. Engineers working on the bridge had to find a way to damp the vibrations and thereby reduce or eliminate the noise.

As with the previous two problems, this problem is a modern version of one that has been in existence for many centuries. The form in which it principally arose in the past was in relation to the sounds made by stringed instruments such as violins and guitars. In these instruments a metal string is stretched between two supports and when the string is displaced by plucking or rubbing it begins to vibrate and emit a musical note.

The frequency of the note is the same as the frequency of the vibrations of the string and so the problem becomes one of relating the frequency of vibration to the properties of the string. In the case of the bridge, the aim is to prevent the vibrations or else to damp them out as quickly as possible when they begin.

The analysis of the vibrations is a complex problem which can be approached in stages, beginning with the simplest possible type of model. Any vibrating system has a natural frequency at which it will vibrate if set in motion. If a force is applied to the system which tries to make it vibrate at this frequency, then the system will vibrate strongly and in some systems catastrophic results can follow if the vibrations are not damped out. An example of this is one of the most famous bridge collapses in history. This occurred on 7 November 1940, when the Tacoma Narrows Bridge in the United States had only been open for a few months. A moderately strong wind started the bridge vibrating with its natural frequency. The results were spectacular. News movies show the entire bridge oscillating wildly in a wave-like motion before it was finally wrenched apart.

Figure 1.6: The Anzac Bridge

With the Anzac Bridge, the vibrations caused annoyance rather than catastrophe, but the problem needed to be dealt with. The first step is to find, at least approximately, the natural frequency of the vibration of the cables supporting the bridge. Once this frequency is known, then measures can be taken to damp the vibrations.

EXERCISES 1.4

1. What factors do you think are significant in determining the frequency of vibration of a string or cable stretched between two supports? Explain how these factors allow a stringed instrument to be tuned.
2. Musical instruments such as guitars, violins, cellos and double basses rely on vibrating strings to produce sound. Explain why a double bass sounds so different to a violin.

1.5 DISCUSSION

The three problems we have outlined are very different, but they have some features in common. In each of them there is a complex system which can be described in terms of a collection of properties. The properties may describe the system itself or its mode of operation. Some of these properties can be assigned numerical values while others cannot. For example, in the three problems we have described, the time taken for the car to fall from D to C, the atmospheric pressure and the frequency at which the cables vibrate are all numerical properties. On the other hand, the amount of terror caused by the ride, the difficulty experienced by passengers in breathing or the beauty of the bridge do not have exact numerical values.

In the mathematical analysis of problems, we will only consider the numerical properties of objects or systems. It often happens that we may wish to calculate some numerical property of a system from a knowledge of other numerical properties of the system. In almost all such cases, calculating a numerical value will involve the concept of a *function*. We will become quite precise about this concept in the next chapter, but for the moment we shall consider a few special cases.

In the *Tower of Terror*, there are a number of possibilities. We may wish to know the height h of the car at any time t, or the velocity v at any time t, or the velocity v at a given height h. Suppose we consider the variation of height with respect to time. We let h be the height above the baseline, which is taken to be the level of C in Figure 1.2. This height changes with time and for each value of t, there is a unique value of h. This is because at a given time the car can only be at one particular height. Or, to put it another way, the car cannot be in two different places at the same time. On the other hand, the car may be at the same height at different times; two different values of t may correspond to the same value of h. The quantities t and h are often referred to as *variables*. Notice also that the two quantities h and t play different roles. It is the value of t which is given in advance and the value of h which is then calculated. We often call t the *independent variable*, since we are free to choose its value independently, while h is called the *dependent variable*, since its value depends on our choice of t.

In the second problem, we have a similar situation. We wish to find a rule which enables us to find the pressure p at a given height h. Here we are free to choose h (as long as it is between 0 and the height of the atmosphere), while p depends on the choice of h, so that p is the dependent variable and h is the independent variable.

Finally, in the case of the Anzac Bridge, we can represent one of the cables schematically as shown in Figure 1.7. In the figure, the cable is fixed under tension between two points A and B. If it is displaced from its equilibrium position, it will vibrate. To find the frequency of the vibration we need a rule which relates the displacement of the center of the cable x to the time t. Here the independent variable is t, the dependent variable is x and we want a computational rule which enables us to calculate x if we are given t.

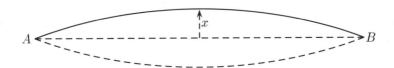

Figure 1.7: A vibrating cable

In a given problem, there is often a natural way of choosing which variable is to be the independent one and which is to be the dependent one, but this may depend on the way in which the problem is posed. For example, in the case of air pressure, we can use a device known as an altimeter to measure height above the earth's surface by observing the air pressure. In this case, pressure would be the independent variable, while height is the dependent variable.

1.6 RULES OF CALCULATION

Each of the above problems suggests that to get the required numerical information, we need a rule of calculation which relates two variables. There are many ways of arriving at such rules—we may use our knowledge of physical processes to deduce a mathematical rule of calculation or we may simply observe events and come to trial and error deductions about the nature of the rule.

Let us try to distil the essential features of rules of calculation which can be deduced from the three examples we have presented:

- The independent variable may be restricted to a certain range of numbers. For instance, in the case of the *Tower of Terror* we might not be interested in considering values of time t before the motion of the car begins or after it ends. In the case of atmospheric pressure, the height must not be less than zero, nor should it extend to regions where there is no longer any atmosphere. This range of allowed values of the independent variable will be called the *domain* of the rule.

- There has to be a procedure which enables us to calculate a value of the dependent variable for each allowable value of the independent variable.

- To each value of the independent variable in the domain, we get one and only one value of the dependent variable.

In the next chapter, we shall formalise these ideas into the concept of a *function*, one of the most important ideas in mathematics.

EXERCISES 1.6

1. The pressure on the hull of a submarine at a depth h is p. Explain how we can regard either of the variables p and h as the independent variable and the other as the dependent variable. Suggest a reasonable domain when h is the independent variable.

2. Hypothetical data values for the *Tower of Terror* are given in the table below, where h is the height at time t.

t	0	1	2	3	4	5	6	7	8	9
h	15.0	54.1	83.4	102.9	112.6	112.5	102.6	82.9	53.4	14.1

    Suggest reasons why we have to take t as the independent variable, rather than h. What is a reasonable domain for t?

3. Hypothetical census data for the population of a country region is given in the table below.

Year	1950	1960	1970	1980	1990
Population(Millions)	10	12	14	16	18

    Decide on suitable independent and dependent variables for this problem. Is there only one possible choice? Plot a graph of the data and use the graph to predict the population in 2000. Can you be confident that your prediction is correct? Explain.

4. In the text we stated that the amount of terror experienced by the passengers in the *Tower of Terror* could not be assigned an exact numerical value. However, some people may be more terrified than others, so there is clearly something to measure even if this can't be done exactly. List three ways you could measure a variable such as terror and invent a function that uses terror as an independent variable.

CHAPTER 2
FUNCTIONS

In this chapter we will give a detailed discussion of the concept of a function, which we briefly introduced in Chapter 1. As we have indicated, a function is a *rule* or *calculating procedure* for determining numerical values. However, the nature of the real world may impose restrictions on the type of rule allowed.

2.1 RULES OF CALCULATION

In the problems we shall consider, we require rules of calculation which operate on numbers to produce other numbers. The number on which such a rule operates is called the *input number*, while the number produced by applying the rule is called the *output number*. Let us denote the input number by x, the rule by letters such as $R, S, \ldots$ and the corresponding output numbers by $R(x), S(x), \ldots$. Thus a rule R operates on the input number x to produce the output number $R(x)$. We can illustrate this idea schematically in the diagram below.

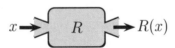

Here we think of a function as a *machine* into which we enter the input number x. The machine then produces the output $R(x)$ according to the rule given by R. A concrete example of such a machine is the common calculator.

EXAMPLE 2.1
Let R be the rule which instructs us to square the input number and then multiply the result by 4.9. In symbols we write $R(x) = 4.9x^2$. Thus, if the input number is 3.1, then its square is 9.61 and multiplying this by 4.9 gives the output number as 47.089.

EXAMPLE 2.2
Let S be the rule which instructs us to find a number whose square is the input number. In symbols, $S(x)$ is given by $(S(x))^2 = x$. Thus if the input number is 4, there are two possible values for the output number $S(4)$, namely -2 and 2. □

As simple as these examples appear, they nevertheless raise points which need clarification. In the three problems that we considered in Chapter 1, we remarked that we needed rules of calculation to compute values of the dependent variables. In these problems, we take the value of the independent variable as the input number for the rule in order to generate the value of the dependent variable or output number. These problems always had a unique value of the dependent variable (output number) for each value of the independent variable (input number). This is certainly not the case for Example 2.2 above, so it seems that not all rules of calculation will be appropriate in practical problem solving.

The second point about the above examples is the fact that there may or may not be restrictions on the values of the input numbers. In the case of Example 2.1, any number can be used as the input number, while in Example 2.2 negative input numbers will not produce an output number. The *natural domain* of a rule is the largest set of numbers which produce an output number. Every rule has a natural domain and to be a solution to a practical problem, a rule must have the property that every physically reasonable value of the independent variable is in the natural domain.

It is sometimes useful to consider a rule in which the set of allowable values of the input number is smaller than the natural domain. The new rule is called a *restriction* of the original rule and such restrictions may have properties not possessed by the original rule. The following examples illustrate some particular cases.

EXAMPLE 2.3
Suppose that a particle moves so that its height h above the earth's surface at time t is given by $h = 25t - 5t^2$. Here t is the input number or independent variable, while h is the output number or dependent variable. The natural domain of the independent variable t is the set of all real numbers. However, if $t < 0$ or if $t > 5$, then h is negative and in the context of this problem, a negative height cannot occur. The physical interpretation of the problem is that the particle begins rising at time $t = 0$, reaches a maximum height before starting to fall and finally reaches the ground again at time $t = 5$. In these circumstances, it is sensible to restrict t to the values $0 \leqslant t \leqslant 5$.

EXAMPLE 2.4
Consider the rule R of Example 2.1 given by $R(x) = 4.9x^2$. Its natural domain is the set of all real numbers. Let P be the same rule, but restricted to the set of positive real numbers: in symbols $P(x) = 4.9x^2$, $x > 0$. For R, each output number is produced by two input numbers. For example, 19.6 is produced by 2 and -2. For P, however, each output number is produced by just one input number. The only input number for the output number 19.6 is 2. □

With these considerations, we are now able to give a precise definition of what we mean by a *function*.

> **DEFINITION 2.1 Functions**
> Let A be a set of numbers. A **function** f with **domain** A is a rule or computational procedure which enables us to calculate a single output number $f(x)$ for each input number x in the set A.

The set of all possible output numbers from a function is called the *range* of the function. It is often quite difficult to determine the range.

You should think carefully about the meanings of the various terms in the definition of a function given above: x is a number, called the *input number* or *independent variable*, f is a rule for calculating another number from x and $f(x)$ is the number we get when we apply this rule to x. We call $f(x)$ the *output number* for x or the *function value* at x. Notice that there is nothing significant about using the letter f to denote a function or the letter x for the independent variable. A function is some rule of calculation and as long as we understand what the rule is, it doesn't matter what letter we use to refer to the function. We can even specify a function without using such letters at all. We simply show the correspondence between the input number and the output number. Thus the function R of Example 2.1 defined by $R(x) = 4.9x^2$ may be written as $x \mapsto 4.9x^2$. We read this as *x goes to (or maps to) 4.9x²*. Common letters to denote functions are f, g and h. Various Greek letters such as ϕ, ψ or δ are also used. The independent variable is a number such as 2 or 4.1237, and it is irrelevant whether we denote it by t, x or any other letter.

EXAMPLE 2.5
We define a function f by the rule
$$f(t) = 2t + 3t^2, \ t \in \mathbb{R}.$$

We also define a function ϕ by the rule
$$\phi(x) = 2x + 3x^2, \ x \in \mathbb{R}.$$

You should convince yourself that f and ϕ define the same function. □

In order to completely specify a function, it is necessary to give both the domain and the rule for calculating function values from numbers in the domain. In practice, it is common to give only the rule of calculation without specifying the domain. In this case it is *assumed* that the domain is the set of all numbers which produce an output number when the rule of calculation is applied. This is called the *natural domain* of the function.

EXAMPLE 2.6
We define a function f by the rule
$$f(t) = \frac{1}{t},$$
where t can be any real number. We can compute $f(t)$ for all $t \neq 0$, so that the natural domain of the function is the set of all numbers t for which either $-\infty < t < 0$ or $0 < t < \infty$. Its range is the same set of numbers.

EXAMPLE 2.7
Let $f(x) = 2x^2 + x + 1$. Find $f(3), f(-x)$ and $f(2x)$.
Solution
$$\begin{aligned} f(3) &= 2(3)^2 + 3 + 1 \\ &= 22 \\ f(-x) &= 2(-x)^2 + (-x) + 1 \\ &= 2x^2 - x + 1 \\ f(2x) &= 2(2x)^2 + (2x) + 1 \\ &= 8x^2 + 2x + 1 \end{aligned}$$

EXERCISES 2.1

In Exercises 1–7, find the numerical value of the function at the given values of a.

1. $f(x) = x^2 - 3$; $a = 3, 5$

2. $f(x) = \dfrac{x-2}{x^2+4}$; $a = 3$

3. $g(x) = \begin{cases} 0 & \text{if } x \leqslant 0 \\ 1 & \text{if } x > 0 \end{cases}$; $a = -1, 4, 5$

4. $f(t) = \dfrac{1}{3-t}$; $a = t^2$

5. $h(x) = 1 + \dfrac{1}{(1+x)^2}$; $a = -3$

6. $\phi(y) = 1 + |2y - 6|$; $a = 2$

7. $g(x) = x^3 - \dfrac{1}{17}x + 2$; $a = 3.23$

In Exercises 8–12, calculate $f(-x), f(1/x)$ and $f(x+1)$.

8. $f(x) = x^2 + 3$

9. $f(x) = \dfrac{1}{3}$

10. $f(x) = \dfrac{x+1}{x^2+1}$

11. $f(x) = x + 1$

12. $f(x) = \begin{cases} 4x + 1 & \text{if } x \leqslant 2 \\ 2x^3 - 7 & \text{if } x > 2 \end{cases}$

13. In Figure 1.2 on page 3, let v be the velocity of the carriage at time t and let h be the vertical height above B at time t. The total energy E at time t is defined to be

$$E = \tfrac{1}{2}mv^2 + mgh,$$

where $g = 9.81 \text{ m/s}^2$ is a constant and m is the mass of the carriage. There is a physical law, known as the *principle of conservation of energy*, which states that E is a constant.

   (a) If the velocity at B is 162 km/hr, what is the velocity at B in m/s?

   (b) Suppose the vertical distance of C from B is 14 m. Use the principle of conservation of energy to calculate the vertical distance CD.

14. Do you think that it is meaningful to consider physical quantities for which no method of measurement is known or given?

15. The population P (in millions) of a city is given by a function f whose input number is the time elapsed since the city was founded in 1876. We have $P = f(t)$. Explain the meaning of the statement $P(107) = 24$.

16. Functions f, g and h are defined by the rules

$$f(x) = 3x, \qquad g(x) = \dfrac{3x^3 + 3x}{x^2 + 1}, \qquad h(x) = \dfrac{3x^3 - 3x}{x^2 - 1}.$$

Explain why $f = g$, but $f \neq h$, $g \neq h$.

17. Express the distance between the origin and an arbitrary point (x_0, y_0) on the line $x + 3y = 1$ in terms of x_0.

18. Let $g(x) = |x - 1|$. When does $g(x)$ equal $x - 1$ and when does it equal $1 - x$?

19. Express the following statements in mathematical terms by identifying a function and its rule of calculation.

    (a) The number of motor vehicles in a city is proportional to the population.

    (b) The kinetic energy of a particle is proportional to the square of its velocity.

    (c) The surface area of a sphere is proportional to the square of its radius.

    (d) The gravitational force between two bodies is proportional to the product of their masses M and m and inversely proportional to the square of the distance r between them.

20. *Challenge problem*: The following function M is defined for all positive integers n and known as McCarthy's 91 function.

$$M(n) = \begin{cases} n - 10, & \text{if } n > 100, \\ M(M(n + 11)) & \text{if } n \leqslant 100. \end{cases}$$

Show that $M(n) = 91$ for all positive integers $n \leqslant 101$.

2.2 INTERVALS ON THE REAL LINE

The domain of a function is often an interval or set of intervals and it is useful to have a notation for describing intervals.

- The *closed interval* $[a, b]$ is the set of numbers x satisfying $a \leqslant x \leqslant b$.
- The *open interval* (a, b) is the set of numbers x satisfying $a < x < b$.
- $(a, b]$ denotes the set of numbers x satisfying $a < x \leqslant b$.
- $[a, b)$ denotes the set of numbers x satisfying $a \leqslant x < b$.

We also need to consider so-called *infinite intervals*.

- The notation $[a, \infty)$ denotes the set of all numbers $x \geqslant a$, while $(-\infty, a]$ denotes the set of all numbers $x \leqslant a$.
- The notation (a, ∞) denotes the set of all numbers $x > a$, while $(-\infty, a)$ denotes the set of all numbers $x < a$.
- We can also use $(-\infty, \infty)$ to denote the set of all real numbers, but this is usually denoted by the special symbol $\mathbb{R}$.

A little set notation is also useful. Let I be an interval on the real line. If x is a number in I, then we write $x \in I$. This is read as x *is in* I, or x *is an element of* I.

Next, let I_1 and I_2 be any two intervals on the real line. Then $I_1 \cup I_2$ denotes the set of all numbers x for which $x \in I_1$ or $x \in I_2$. Note that the mathematical use of the word "or" is not exclusive. It also allows x to be an element of *both* I_1 and I_2. We call $I_1 \cup I_2$ the *union* of I_1 and I_2.

We use $I_1 \cap I_2$ to denote the set of all numbers x for $x \in I_1$ and $x \in I_2$. We call $I_1 \cap I_2$ the *intersection* of I_1 and I_2. It may happen that I_1 and I_2 have no elements in common, in which case the intersection of I_1 and I_2 is said to be *empty*. We write $I_1 \cap I_2 = \emptyset$ in this case.

Finally, we use the notation $I_1 \setminus I_2$ to denote the elements of I_1 which are not also in I_2.

EXAMPLE 2.8
For the function $R(x) = 4.9x^2$ of Example 2.1, the domain is $(-\infty, \infty)$, while the range is $[0, \infty)$. For the function $P(x) = 4.9x^2$, $x > 0$ of Example 2.4, both the domain and range are $(0, \infty)$.

EXAMPLE 2.9
Let $I_1 = [1, 4)$ and $I_2 = (3, 5)$. Then $I_1 \cup I_2 = [1, 5)$, $I_1 \cap I_2 = (3, 4)$ and $I_1 \setminus I_2 = [1, 3]$.

EXAMPLE 2.10
The natural domain of the function
$$f(x) = \frac{1}{x}$$
is $(-\infty, 0) \cup (0, \infty)$.

EXERCISES 2.2

In Exercises 1–6, find the domain and range of the function. Express your answer in the interval notation introduced in this section.

1. $f(x) = |x|$
2. $f(x) = \dfrac{1}{x-1}$
3. $f(x) = \sqrt{1 - 2x}$
4. $f(x) = \dfrac{x}{(x-3)(x+1)}$
5. $g(x) = x^2$, $\quad 0 \leqslant x < 1$
6. $h(x) = \sqrt{x+3}$, $\quad x \geqslant 0$

7. Express the area of a circle as a function of its circumference, that is, find a function f for which the output number $f(C)$ is the area whenever the input number C is the circumference. What is the domain of the function?
8. Express the area of an equilateral triangle as a function of the length of one of its sides.

Rewrite the expressions in Exercises 9–14 as inequalities.

9. $x \in (5, 8]$
10. $x \in (1, \infty)$
11. $x \in (1, 2)$
12. $x \in (1, 9] \cup [10, \infty)$
13. $x \in (-1, 4) \cap (2, 6)$
14. $x \in (-\infty, \infty)$

Rewrite the following expressions in interval notation.

15. $2 \leqslant x < 3$
16. $x > \pi$
17. $x \in \mathbb{R}$
18. $x \geqslant 12$
19. $|x + 3| < 4$
20. $|x| \geqslant 2$
21. $-1 < x < 3$ and $x > 5$

2.3 GRAPHS OF FUNCTIONS

If f is a function with domain D, then its *graph* is the set of all points of the form $(x, f(x))$, where x is any number in D (written $x \in D$). It is often difficult to sketch graphs, but one method is simply to plot points until we can get an idea of the nature of the graph and then join these points, as we have done in Example 2.11 below. The procedure for any other example is no different—it may just take longer to compute the function values. In many cases there are shorter and more elegant ways to plot the graph, but we shall not investigate these methods in any detail.

EXAMPLE 2.11
Let the function f be defined by the rule $f(x) = x^2$. We tabulate values of $f(x)$ for various values of x:

x	-4	-3	-2	-1	0	1	2	3	4
$f(x)$	16	9	4	1	0	1	4	9	16

If these points are now plotted on a diagram the general trend is immediately evident and we can join the points with a smooth curve. This is shown schematically in the diagram below, which we have plotted using the software package *Mathematica*.

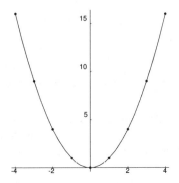

Figure 2.1: The graph of $f(x) = x^2$

□

Why do we draw graphs? One of the main reasons is as an aid to understanding. It is often easier to interpret information if it is presented visually, rather than as a formula or in tabular form. With the advent of computer software such as *Mathematica* and *Maple*, the need to plot graphs by hand is not as great as it used to be. Computers plot graphs in a similar way to the above example—they calculate many function values and then join neighbouring points with straight lines. Because the plotted points are so close together, the straight line segments joining them are very short, and the overall impression we get from looking at the graph is that a curve has been drawn.

2.3.1 Plotting graphs with *Mathematica*

The instruction for plotting graphs with *Mathematica* is **Plot**. The essential things that *Mathematica* needs to know are the function to be plotted, the independent variable and the range of values for the independent variable. There are numerous optional extras such as colour, axis labels, frames and so on. In this book we will explain how to use *Mathematica* to perform certain tasks, but we will assume that you are familiar with the basics of *Mathematica*, or have access to supplementary material.[1]

EXAMPLE 2.12
To plot a graph of the function given by $f(x) = x^2$, we use the instruction

```
Plot[x^2,{x,-4,4}]
```

This will produce the following graph:

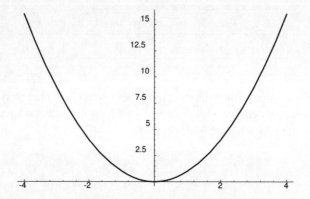

Figure 2.2: A basic *Mathematica* plot

This graph does not look as nice as the one in Example 2.1. The vertical axis is too crowded, the curve is rather squashed and there is no colour. To improve the appearance, we can add a few more options. To plot the graph in Example 2.1 (without the dot points), we used the instruction

```
Plot[x^2,{x,-4,4},PlotStyle->{RGBColor[1,0,0]},
Ticks->{{-4,-2,0,2,4},{5,10,15,20}},AspectRatio->1]
```

Here the `Ticks` option selects the numbers that we wish to see on the axes, while `AspectRatio` alters the ratio of the scale on the two axes. Notice that *Mathematica* may sometimes choose to override our instructions. In this case it has not put a tick for $y = 20$. The numbers in `RGBColor` gives the ratios of red, green and blue. In this case the graph is 100% red.[2]   □

EXERCISES 2.3

1. Use the *Mathematica* instruction given in the text to plot the graph of $f(x) = x^2$. Experiment with different colours and aspect ratios to see how the appearance of the graph changes.

---

[1] For example, *Introduction to Mathematica* by G. J. McLelland, University of Technology, Sydney, 1996.
[2] The current edition of this book has not been printed in colour. However, you can get colour graphs on a computer by following the given commands.

2. Plotting a graph over a part of its domain can lead to deceptive conclusions about the nature of the graph in its entire domain. By selecting a suitable interval, obtain the following plots for $f(x) = x^6$ and $g(x) = 2^x$. In each case, the dotted plot is the graph of g. What would you have concluded had you plotted only one of these graphs?

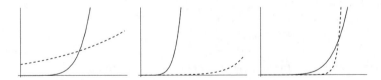

3. Here is a table of values which could apply to the *Tower of Terror*.

t	0	1	2	3	4	5	6	7	8	9
h	15.0	54.1	83.4	102.9	112.6	112.5	102.6	82.9	53.4	14.1

(a) Use the command

a={{0,15},{1,54.1},{2,83.4},{3,102.9},{4,112.6},
{5,112.5},{6,102.6},{7,82.9},{8,53.4},{9,14.1}}

to enter this table as a list in a *Mathematica* notebook.

(b) Use the *Mathematica* command ListPlot in the form

g0=ListPlot[a,PlotStyle->PointSize[0.02]]

to plot these points on a graph. Notice that these points lie on what appears to be a smooth cur Print a copy of the graph, join the points by hand with a smooth curve and use this curve estimate the height after 2.5 s.

(c) Rather than estimate the curve by joining the points by hand, we can use the *Mathematica* command Fit. This command fits a curve of the user's choice (straight line, quadratic polynomial and so on) to a list of data points. There are numerous options and you should consult the *Mathematica* help files for a full explanation of this command. The command

f1=Fit[a,{1,x},x]

will give a formula for the best straight line fitting the given points. Similarly, the commands

f2=Fit[a,{1,x,x^2},x]
f3=Fit[a,{1,x,x^2,x^3},x]

will give a formula for the best quadratic and best cubic polynomial respectively fitting the giv points. Fit a quadratic polynomial to the points in the above table, that is, find f2. Use t formula to calculate the height at each of the times given in the table. How good is the agreeme What conclusions can you draw about the data? Determine the height after 2.5 seconds. D this agree with your answer in part (b)?

(d) You can plot the formula f2 with the instruction

$$g2=\text{Plot}[f2,\{x,0,10\}]$$

Do this, and then display the graph g2 and the plotted points g0 simultaneously by using the instruction

$$\text{Show}[g0,g2]$$

4. Here is a table giving the atmospheric pressure at different heights:

h	0	1000	2000	3000	4000	5000	6000	7000	8000	9000	10 000
p	1013	899	795	701	616	540	472	411	356	307	264

(a) Enter this table as a list in a *Mathematica* notebook

(b) Plot these points on a graph. Notice that these points lie on what appears to be a smooth curve. Print a copy of the graph, join the points with a smooth curve and use this curve to estimate the pressure at a height of 4500 metres.

(c) Fit a straight line, a quadratic polynomial and a cubic polynomial to the set of points in the table above. In each case, use the formula to calculate the pressure at the heights given in the table and compare the results with the tabulated values. Which curve gives the best approximation to the table values? Compute the pressure at a height of 4500 m and compare it with your result in part (b).

(d) On the same set of axes, plot the graphs of the functions obtained in part (c).

5. A airplane flying from Tullamarine Airport in Melbourne to Kingsford-Smith in Sydney has to circle Kingsford-Smith several times before landing. Plot a graph of the distance of the airplane from Melbourne against time from the moment of takeoff to the moment of landing.

6. A rectangle of height x is inscribed in a circle of radius r. Find an expression for the area of the rectangle. Plot a graph of this area as a function of x and decide from your graph at what value of x the area of the rectangle is a maximum.

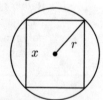

2.4 EXAMPLES OF FUNCTIONS

In this section, we consider various examples of functions. Looking at different examples is a useful learning method in mathematics and one which you should try to cultivate. Very often, mathematical concepts possess subtleties which are not immediately apparent. Studying and doing many examples exposes you to many different aspects of a concept and should help your learning. Before giving these examples, there is one matter which needs to be clarified, and that is the one posed by the following question:

What do we mean when we say that a given function f, say, is *known* or *well-defined*?

In this book, we emphasise the use of functions to compute numerical data relating to problems in the real world. This implies that when we use a function to calculate the value of the dependent variable in a practical problem, the answers we get should agree with the experimental data. As far as such data is concerned, the only values which are measured in the real world are decimal numbers to a certain accuracy. Since more sophisticated methods of measurement may increase accuracy, the rule that defines a function should be able to produce output numbers in decimal form with arbitrary accuracy. If the rule does not allow us to do this, then we cannot really say that the function is known. Accordingly, one answer to the question posed above is:

> We say that the function f is *known* or, more technically, is *well-defined*, if we are able to compute $f(x)$ in decimal form to an arbitrary degree of accuracy, for any x in the domain of f.

For example, the rule R of Example 2.1 given by $R(x) = 4.9x^2$ enables us to compute $R(x)$ to any number of decimal places that we choose. Computing an output number such as $R(1626.876509)$ to any degree of accuracy can be done, although it may take some time.[3] Thus R is a well-defined function.

Now let us turn to some examples. In describing any function, we must insist on three things:

1. The domain of the function must be known. (But see the remark on page 13 about natural domains.)

2. There must be a rule which produces one (and *only* one) output number for each input number in the domain.

3. The rule must be in a form which enables us to compute the output number to any specified degree of accuracy.

One simple way of defining functions is by arithmetic operations—addition, subtraction, multiplication and division, and the simplest functions of this type are *polynomials*. These consist of sums of the form
$$f(x) = a_n x^n + a_{n-1} x^{n-1} + \cdots + a_1 x + a_0,$$
where $a_0, a_1, \ldots, a_n$ are constants for a given f.

EXAMPLE 2.13
A function f defined by the rule
$$f(x) = x^3 - 4x^2 + x + 6$$
is a polynomial function. We can compute its value for any input number, for example
$$\begin{aligned} f(2.1) &= 2.1^3 - 4 \times 2.1^2 + 2.1 + 6 \\ &= 9.261 - 17.64 + 2.1 + 6 \\ &= -0.279. \end{aligned}$$

The (natural) domain of the function is $\mathbb{R}$ and it can be shown that the range is also $\mathbb{R}$. We can plot the graph either by plotting points by hand and joining them or, more conveniently, by using a graphics package such as *Mathematica*. The graph of the function is shown in Figure 2.3. □

A diagram such as Figure 2.3 may give a good idea of the appearance of the graph, but it does not tell us certain critical properties. For example, are there any turning points of the graph besides the

---
[3] The exact answer is 12968963.1601265326969, obtained using *Mathematica*.

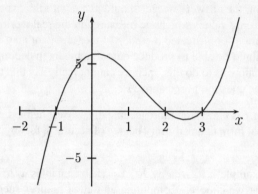

Figure 2.3: The graph of $f(x) = x^3 - 4x^2 + x + 6$

ones shown? In many cases of interest, the domain of a function is an infinite interval, in which case we can only plot part of its graph. We cannot then be sure about features, such as turning points, that may occur in the region we have not plotted. In Figure 2.3 all the turning points of the graph *have* been plotted, but we would need to do some further mathematical analysis to show this.

In the above example, we have specified the rule for calculating function values by giving a formula which applies at all points of the domain. However, the rule for calculating function values may be given by a different formula at different points of the domain, as the following example shows.

EXAMPLE 2.14
We define a function ϕ by the following rule:

1. The domain of ϕ is $\mathbb{R}$.

2. If x is the input number, then:

   - The output number is 1 if $x < 0$.
   - The output number is $2x + 1$ if $0 \leqslant x < 1$.
   - The output number is $3(x - 1)$ if $1 \leqslant x < 2$.
   - The output number is 3 if $x \geqslant 2$.

In symbols we can write
$$\phi(x) = \begin{cases} 1 & \text{if } x < 0, \\ 2x + 1 & \text{if } 0 \leqslant x < 1, \\ 3(x - 1) & \text{if } 1 \leqslant x < 2, \\ 3 & \text{if } x \geqslant 2. \end{cases}$$

As we have already mentioned, the domain of ϕ is $\mathbb{R}$. The rule for computing function values obviously produces numerical values of any accuracy:

$$\phi(1.75) = 3(1.75 - 1)$$
$$= 2.25$$
$$\phi(2.5) = 3$$
$$\phi(-1.5) = 1$$

The graph of the function is shown in Figure 2.4. From the graph we see that the range is $[0, 3]$. Notice the small circle centred at the point $(1, 3)$. Here we are using the common convention that this circle indicates that the point at its centre—in this case $(1, 3)$—is *excluded* from the graph. We have also centered a solid disk at the point $(1, 0)$ to indicate that this point is *included* in the graph.

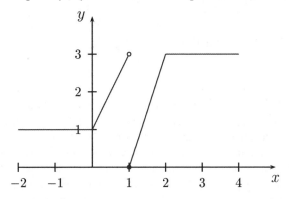

Figure 2.4: The graph of $y = \phi(x)$

EXAMPLE 2.15

Allowing the operation of division in the rule produces more complicated examples. Let us define a function f by

$$f(x) = \frac{x^2 - x - 2}{x^2 + x - 2}.$$

Notice that the denominator is zero for $x = -2$ and $x = 1$, so these numbers are excluded from the natural domain. Since we can use any other number in the above expression for $f(x)$, the natural domain is the set of all real numbers other than -2 and 1. This is written is various ways, such as $\mathbb{R} \setminus \{-2, 1\}$ or $(-\infty, -2) \cup (-2, 1) \cup (1, \infty)$. Here $\{-2, 1\}$ denotes the set consisting of the two elements -2 and 1, while $\mathbb{R} \setminus \{-2, 1\}$ denotes the set $\mathbb{R}$ with the two points -2 and 1 removed.

Numerical evaluations of the function values can be carried out by hand, but they become tedious:

$$f(1.5) = \frac{2.25 - 1.5 - 2}{2.25 + 1.5 - 2}$$
$$= -\frac{1.25}{1.75} = -\frac{5}{7}$$

To sketch the graph, it is easier to let *Mathematica* do all the work of evaluating function values and plotting the curve. The result is shown in Figure 2.5.

The dotted lines $x = -2$ and $x = 1$ are not part of the graph. They are called vertical *asymptotes* and we had to give *Mathematica* special instructions in order to include them on the diagram. □

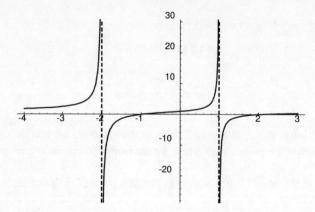

Figure 2.5: A *Mathematica* plot of the graph of $f(x) = \dfrac{x^2 - x - 2}{x^2 + x - 2}$

Functions defined by algebraic formulas arise in some simple applications, but in most applications the functions we obtain as solutions have to be defined in other ways. We give an example of such a function, whose rule of evaluation may seem strange, artificially contrived and of no practical use. In fact the rule is very useful in applications and a large part of our later work will be to find out why such rules are needed in practical problems and how we can obtain them. It would be very easy to solve problems if the only functions we ever needed were nice simple ones. Unfortunately (or perhaps fortunately, depending on your point of view), describing phenomena in the real world is usually a difficult procedure and if we are to make any progress we need some non-trivial mathematics.

EXAMPLE 2.16

We are going to define a function E by a rule which is, at first sight, unusual. If you have not seen this rule before, you probably will have no idea of what the function is used for or where the rule comes from. The purpose of this example is to show that no matter how complex or unusual a rule may seem, if it is properly specified we should be able to follow it and produce an output number. At this point we are not going to explain what the function E is used for, but only how to calculate its value for any input number. We are asking you to go through the same sort of unthinking process that a computer uses in calculations—just follow the rules. The *why* of this example will be clarified in Chapter 6. Here is the process:

1. Let the input number x be any positive real number.

2. Select a desired accuracy of calculation ε, that is, we want the calculation to give an output number with an accuracy of $\pm\varepsilon$.

3. Calculate each of the numbers
$$1, x, \frac{x^2}{2!}, \frac{x^3}{3!}, \frac{x^4}{4!}, \ldots,$$
where for any $k \geqslant 1$, we define $k! = k \times (k-1) \times (k-2) \times \cdots \times 3 \times 2 \times 1$. Continue until one of the numbers is less than ε. Suppose this number is $\dfrac{x^n}{n!}$.

4. Then the output number for x can be calculated with an error of less than ε from the expression
$$E_n(x) = 1 - x + \frac{x^2}{2!} - \frac{x^3}{3!} + \cdots + (-1)^{n-1}\frac{x^{n-1}}{(n-1)!},$$

that is, $E_n(x) - \varepsilon < E(x) < E_n(x) + \varepsilon$.

As an illustration, take $x = 1.5$ and $\varepsilon = 0.001$. Then we find

$$x = 1.5 \qquad \frac{x^2}{2!} = 1.125 \qquad \frac{x^3}{3!} = 0.5625 \qquad \frac{x^4}{4!} \doteq 0.2109$$
$$\frac{x^5}{5!} \doteq 0.0633 \qquad \frac{x^6}{6!} \doteq 0.0158 \qquad \frac{x^7}{7!} \doteq 0.0034 \qquad \frac{x^8}{8!} \doteq 0.0006,$$

where we have worked to 4 decimal places. The notation $\doteq$ is used to indicate that the number on the right hand side is, to the stated number of decimal places, equal to the number on the left hand side. Following our rule, we get

$$E(1.5) \doteq 1 - 1.5 + 1.125 - 0.5625 + 0.2109 - 0.0633 + 0.0158 - .0034$$
$$= 0.2225$$
$$\doteq 0.223.$$

We conclude that $0.222 \leqslant f(1.5) \leqslant 0.224$. To 2 decimal places we have $E(1.5) \doteq 0.22$, a number we can plot on a graph. If we were to repeat this calculation for many input numbers, plot them, and interpolate a smooth curve, we get the graph of $E(x)$ shown in Figure 2.6. □

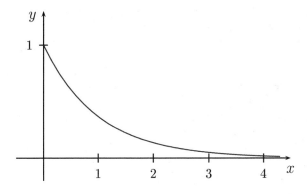

Figure 2.6: The graph of $y = E(x)$

The message of this example is that if a function is properly defined, then we can compute its values, even if we have no idea where the function comes from or what it is used for. When we program a computer we have to engage in a similar process, that is, carefully specify the rules that enable the computer to calculate function values.

We will see in Chapter 6 that a function of this type is needed to produce the solution to the air pressure problem we considered in Section 1.3.

Finally, let us turn to the problem of finding the range of a function. As we have mentioned earlier, finding the range of a function can be difficult. In the case of rational functions with quadratic terms, we can sometimes use our knowledge of quadratic equations to find the range. A *rational function* is one that can be expressed as the quotient of two polynomials.

EXAMPLE 2.17
Find the range and natural domain of the function

$$f(x) = \frac{x}{x^2 - 4}.$$

It is easy to see that the natural domain is the set of all real numbers x for which $x \neq \pm 2$. To find the range, we put

$$y = \frac{x}{x^2 - 4}$$

and rearrange the equation to get $yx^2 - x - 4y = 0$. This is a quadratic equation in x and we need to find the values of y for which this equation can be solved to give real values of x. The discriminant is

$$\Delta = 1 + 16y^2$$

and this is well-defined for all y. Hence the range of f is $\mathbb{R}$.

EXERCISES 2.4

In Exercises 1–6, find the natural domain and range of the function.

1. $f(x) = x^2 - 3$
2. $f(x) = \dfrac{x + 1}{x^2 - 4}$
3. $g(t) = \sqrt{t^2 - 1}$
4. $x(f) = f^2 - 3$
5. $g(x) = x + \dfrac{1}{x}$
6. $f(x) = x^3 - x$

7. Let f be the function defined in Example 2.16. In this example, we found $f(1.5) = 0.223 \pm 0.001$. Evaluate $f(0)$, $f(0.5)$, $f(1)$ and $f(2)$ with an error of less than 0.001. Plot these values on a graph, together with the value of $f(1.5)$. Hence verify that the graph of f is consistent with the one given in Figure 2.6.

8. Consider the equation

$$x^2 + y^2 = 1. \qquad (*)$$

For every value of x, we can find two possible values of y, namely $y = \sqrt{1 - x^2}$ and $y = -\sqrt{1 - x^2}$. Thus, each of the above equations defines a function of x according to

$$f_1(x) = \sqrt{1 - x^2}$$

and

$$f_2(x) = -\sqrt{1 - x^2}.$$

We say that f_1 and f_2 are functions *implicitly* defined by the equation $(*)$. We can also solve the equation $(*)$ for x in terms of y, but this does not give rise to new functions: the function $g_1(y) = \sqrt{1 - y^2}$ is the same function as f_1. However, the functions f_1 and f_2 are not the only functions implicitly defined by equation $(*)$. Can you think of some others?

9. Can you give a computational procedure for calculating square roots without using tables or built-in calculator functions? If your answer is "no", does this mean that you don't really understand what is meant by an expression such as $\sqrt{2}$? Comment. (The matter of computing square roots will be taken up in Chapter 10.)

CHAPTER 3
CONTINUITY AND SMOOTHNESS

In this book we are mainly interested in functions which provide the solutions to mathematical models of real situations. In Chapter 2, we gave some examples of functions. However, not all of these will occur in practical problems. A function such as that of Figure 2.5 would rarely, if ever, occur in a practical problem. The function shown in Figure 2.4 may also seem to be in this category, but in fact functions of this type have many uses, although we shall not consider them in any detail in this book.

In general, the processes which occur in the real world happen smoothly. We do not expect to see bodies jump from one place to another in no time or to instantaneously change their velocity. We would not expect the air pressure at a particular height to change suddenly from one value to a completely different one. There are of course processes which do occur suddenly, such as switching on a light, hitting an object with a hammer or perhaps falling off a cliff, but these are really smooth processes which occur very rapidly on our time scale. The functions which occur in the problems in this book will all be smooth.

3.1 SMOOTH FUNCTIONS

What does it mean for a function to be smooth? In order to reach an answer to this question, let us consider some particular cases.

Suppose a body moves along a straight line with its displacement s from a fixed point O at time t being given by $s = f(t)$ for some function f. What can we say in general terms about the graph of f? Consider the two graphs in Figure 3.1.

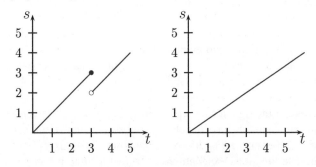

Figure 3.1: Displacement of a moving body

28 CONTINUITY AND SMOOTHNESS

Let us look at the figure on the left. For $t \leqslant 3$ seconds, the graph represents a body which moves away from O, getting closer to a distance of 3 metres from O as t gets closer to 3 seconds. When $t = 3$, the body is 3 units from O. Immediately after 3 seconds have elapsed the body instantaneously moves to a position only 2 metres from O. Clearly, such behaviour is not in accord with our everyday experience.

The graph on the right is more reasonable. Here the distance from O increases without any sudden changes in position. The obvious difference between the two graphs is the jump in the first one. Such difference in behaviour is described by the notion of *continuity*: the graph on the right hand side is continuous, while the one on the left is discontinuous.

Next, suppose a body moves along a straight line so that its displacement s from a fixed point O at time t is given by the graph in Figure 3.2.

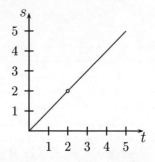

Figure 3.2: Displacement of a moving body

As we have mentioned earlier, the small circle centred at the point $(2, 2)$ indicates that the point at its centre—in this case $(2, 2)$—is excluded from the graph. The physical interpretation of such a graph is that at the time $t = 2$ seconds, the body has no position, because there is no point on the s–axis corresponding to the point $t = 2$ seconds. This behaviour is another example of lack of continuity and is quite unrealistic. We would like to exclude such displacement–time graphs from our considerations.

The essential idea of continuity that emerges from these examples is that the graph of a smooth function should have no gaps or jumps.

There is another aspect to the idea of smoothness, which we illustrate by considering two different cases for the velocity of a moving body. First there is the case where a body moves with constant speed v in a straight line. If s denotes the distance from a fixed point O at time t, then $s = vt$ and a plot of s against t gives a straight line with slope v. The left hand graph of Figure 3.3 shows the case $v = 1$. Now imagine a body which, at time $t = 2$, undergoes an instantaneous change of velocity from 0.5 m/sec to 1 m/sec. The right hand graph of Figure 3.3 shows this case. Up to the time $t = 2$, the graph will be a straight line of slope $\frac{1}{2}$. At $t = 2$, the graph changes instantaneously to a straight line of slope 1. Such behaviour of moving bodies does not occur in the real world and we want to investigate the restrictions needed to rule it out.

The noticeable feature of the right hand graph is the corner at $t = 2$. This is what we have to avoid if we are to rule out instantaneous changes in velocity. Loosely speaking, a function is *differentiable* if its graph has no corners. We often refer to differentiable functions as *smooth* functions. We shall later show that all differentiable functions are continuous, so that a smooth function is one which is both continuous and differentiable. We have at present no way of determining whether or not a function is differentiable, continuous or neither except to look at its graph. However, a graph constructed

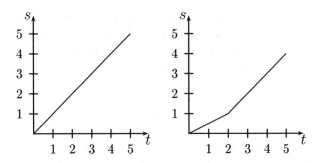

Figure 3.3: Smooth and non-smooth displacements

simply by plotting points is inconclusive, because anything might happen between the plotted points. Figure 3.4 shows an example where the graphs of two functions pass through the same plot points, but one function is smooth while the other is not. To know whether or not a function is smooth, we need methods which will give conclusive answers and to do this, we will need definitions to tie down the exact meaning of the concepts of continuity and differentiability.

There is a second reason for needing these definitions. Our main aim is to find the function which relates the variables in a practical problem. It turns out that a careful look at the concept of smoothness will point the way towards methods for finding such functions.

We should remark that in some applications we need to make use of functions which are not smooth. The income tax scale is one example. There are also cases where smooth functions change in very short time periods. It is often simpler to approximate such smooth functions by a non-smooth function. This may occur when we model the turning of a switch on or off. We will not consider such applications in this book.

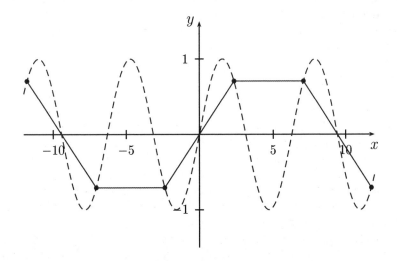

Figure 3.4: Two graphs through the same plot points

As is usual in mathematics, we will have to give precise and unambiguous definitions of the concepts discussed above. Historically, such definitions did not come out of the clear blue sky, or inscribed on stone tablets. Rather, they developed slowly over time in response to some perceived need. Often concepts are taken for granted or are not properly understood, until some example arises which forces us to take more care with their definitions. In our case, there is a need for careful definitions of continuity and differentiability to ensure that the functions we use to describe real world processes are in accordance with experience.

EXERCISES 3.1

1. The square wave function is defined on the interval $[0, 2)$ by

$$\phi(t) = \begin{cases} 1 & \text{if } 0 \leqslant t < 1, \\ 0 & \text{if } 1 \leqslant t < 2. \end{cases}$$

We extend the definition of ϕ to the entire real line by letting

$$\phi(x + 2) = \phi(x)$$

for all $x \in \mathbb{R}$. Sketch the graph of ϕ on the interval $[-6, 6)$.
This function is often used to model a signal generated by repeatedly flicking a switch on and off. Do you think that the function ϕ is an accurate representation of the physical process?

3.2 CONTINUITY

Looking at the graphs of the functions in Figures 2.4, 2.5, 3.1 and 3.2, we see that the jumps or gaps in the graph occur at specific points in the domain and we describe this by saying that continuity is primarily a *point property*. A function may be continuous at some points and not continuous at others. In attempting to formulate a precise definition of continuity, we begin with the idea of *continuity at a point*. There are several criteria—not necessarily independent of each other—which have to be satisfied if a function is to be continuous at some point of its domain. We will motivate such criteria by considering various examples.

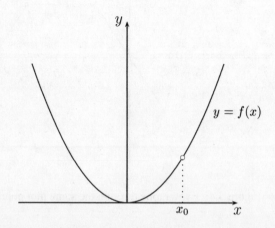

Figure 3.5: A graph with a gap

To begin with, consider the graph of the function f shown in Figure 3.5. In this case, the function is not defined at the point x_0 and this is manifested by the gap which appears in the graph above the point $x = x_0$. Clearly, the only way to avoid having such a gap is to ensure that there is indeed a function value at x_0, that is, to ensure that $f(x_0)$ is well-defined. This gives us our first criterion for continuity.

Criterion 1 In order for a function f to be continuous at a point x_0, the number $f(x_0)$ must exist, that is, x_0 must be in the domain of f.

For our next example, let us consider the function $\phi(x)$ given by

$$\phi(x) = \begin{cases} 1 & \text{if } x < 0, \\ 2x + 1 & \text{if } 0 \leq x < 1, \\ x + 1 & \text{if } 1 \leq x < 2, \\ 3 & \text{if } x \geq 2. \end{cases}$$

The graph of $\phi(x)$ is shown in Figure 3.6. Clearly, we would not want to consider this to be continuous at the point $x = 1$, so that we need a condition that describes this type of jump behaviour. In this

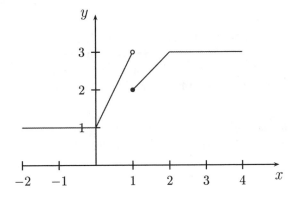

Figure 3.6: A jump discontinuity

example, $f(1) = 2$, while the values of $f(x)$ for x slightly less than 1 are close to 3 and the values of $f(x)$ for x slightly greater than 1 are close to 2. There would be no jump if values of $f(x)$ were close to 2 (the value of $f(1)$) for *all* x close to 1. This gives us another criterion for continuity at a point.

Criterion 2 In order for a function f to be continuous at a point x_0 of its domain, it is necessary that the values of $f(x)$ are close to the value of $f(x_0)$ as long as x is close to x_0.

The word "close" is not very precise: how close is "close"? We can sharpen the idea a little by saying that the values of $f(x)$ should be arbitrarily close to the value of $f(x_0)$ provided x is sufficiently close to x_0. However, in our final definition of continuity we will have to be more careful about the meanings of the various terms. For the moment we will continue at an intuitive level.

We can now state a preliminary definition of continuity at a point.

32 CONTINUITY AND SMOOTHNESS

> **DEFINITION 3.1 Preliminary definition of continuity at a point**
> *The function f is **continuous at the point** x_0 in its domain if:*
>
> 1. *The function f is defined at the point x_0, that is, we can compute the number $f(x_0)$.*
>
> 2. *The values of $f(x)$ are arbitrarily close to the value of $f(x_0)$ as long as x is sufficiently close to x_0.*

The above definition of continuity definition lacks precision, since it makes use of the idea of *sufficiently close*. One way to clarify this is to use the idea of a *sequence*. Loosely speaking, a sequence is an infinite list of numbers. Some examples of sequences are given below.

EXAMPLE 3.1
 (a) $1, 4, 9, 16, \ldots$

 (b) $1, -1, 1, -1, 1, -1, \ldots$

 (c) $1, \frac{1}{2}, \frac{1}{3}, \frac{1}{4}, \ldots$

 (d) $\left(1 + \frac{1}{1}\right)^1, \left(1 + \frac{1}{2}\right)^2, \left(1 + \frac{1}{3}\right)^3, \left(1 + \frac{1}{4}\right)^4, \ldots$

$\square$

In order for a sequence to be well-defined, there must be a rule for computing its terms. A sequence is often referred to by giving the general term in braces. Thus the above four sequences are $\{n^2\}$, $\{(-1)^n\}$, $\{1/n\}$ and $\{(1+1/n)^n\}$. Notice that in each sequence there is one term for each positive integer and a rule for generating this term from the integer. This leads to a precise definition of a sequence.

> **DEFINITION 3.2** *A **sequence** of real numbers is a function whose domain is the set of positive integers. The values taken by the function are called the **terms** of the sequence.*

EXAMPLE 3.2
Define a function f by
$$f(n) = 1/n,$$
where n is a positive integer. Then
$$f(1) = 1, f(2) = \tfrac{1}{2}, f(3) = \tfrac{1}{3}, \ldots,$$
so f is a sequence and its terms are
$$1, \tfrac{1}{2}, \tfrac{1}{3}, \tfrac{1}{4}, \ldots$$

$\square$

Despite the above definition, a sequence is usually thought of as a list of numbers rather than a function.

The most important property of a sequence is whether or not there is some number x_0 to which the terms get arbitrarily close as n gets larger. If this is the case, then the sequence is said to be *convergent* and the number is called the *limit*. If the sequence is not convergent, then it is said to be *divergent*. There are a number of common ways of indicating that the sequence $\{x_n\}$ converges to x_0:

- We may simply state the fact in words: $\{x_n\}$ *converges to* x_0.

- We may write $x_n \to x_0$ as $n \to \infty$. We read this as x_n *approaches* x_0 *as* n *tends to infinity.*

- A more compact way of writing the statement above is $\lim_{n \to \infty} x_n = x_0$.

In simple cases we can determine whether or not the sequence converges by looking at the list of terms. Thus in Example 3.1, the sequence (a) diverges because the terms get larger without any bound, the sequence (b) diverges because the terms oscillate. In each case, there is no number that the terms get close to as n gets large. The sequence (c) converges with limit 0, since the terms get arbitrarily close to 0 as $n \to \infty$. The sequence (d) is more difficult. It is not obvious what happens from the terms given. If we compute the decimal approximations of the first few terms we get

$$2,\ 2.25,\ 2.37,\ 2.44,\ 2.49,\ \ldots$$

It is still not clear what will happen. There are methods for dealing with such sequences and while these are important for many practical purposes we shall not need them in this book. However, this particular sequence is important. In Theorem 10.5 of Chapter 10 we will show that it is convergent with limit $2.71828\ldots$, a number commonly denoted by the letter e. For the present we shall assume this result without proof, that is, we assume

$$\lim_{n \to \infty} \left\{ \left(1 + \frac{1}{n}\right)^n \right\} = e$$
$$= 2.71828\ldots$$

With the idea of a sequence available, we can improve our understanding of the idea of continuity of a function f at a point x_0 in its domain. If we take any sequence $\{x_n\}$ in the domain of f which converges to x_0 and then consider the corresponding sequence of function values $\{f(x_n)\}$, this latter sequence must converge to $f(x_0)$ if f is continuous at x_0. This is another and more precise way of stating Criterion 2. Notice that we have implicitly assumed that $f(x_0)$ exists, which is required by Criterion 1. This property of convergence must hold for every sequence which approaches x_0, and of course each number in the sequence $\{x_n\}$ must be in the domain of the function, so that $\{f(x_n)\}$ is defined for all n.

> **DEFINITION 3.3 Continuous functions**
> *The function f is **continuous at the point** x_0 in its domain if the sequence $\{f(x_n)\}$ converges to $f(x_0)$ for **every** sequence $\{x_n\}$ in the domain that converges to x_0.*

Notice carefully the words "for every" in this statement. We must consider input sequences on both sides of the point x_0, as well as those which oscillate from side to side. However, to show that a function is not continuous at a point x_0, it is enough to find just one sequence $\{x_n\}$ converging to x_0 such that the sequence $\{f(x_n)\}$ does not converge to $f(x_0)$.

EXAMPLE 3.3

Consider the function $R(x) = 4.9x^2$ of Example 2.1. To investigate the continuity at $x = 2$, say, we need to consider input sequences which converge to 2. One of the simplest is $\{x_n\} = \{2 + 1/n\}$. The terms are

$$3,\ 2\tfrac{1}{2},\ 2\tfrac{1}{3},\ 2\tfrac{1}{4},\ \ldots,$$

which converges to 2 as n tends to infinity. The sequence of function values or output sequence is $\{R(x_n)\} = \{4.9(2 + 1/n)^2\}$. The terms of this output sequence are

$$4.9(2+1)^2, 4.9(2+\tfrac{1}{2})^2, 4.9(2+\tfrac{1}{3})^2 \ldots$$

The terms in brackets converge to 2, so the output sequence converges to 4.9×2^2, which is the function value at 2.

We have shown that there is a sequence $\{x_n\}$ with $\lim_{n\to\infty} x_n = 2$ for which $\lim_{n\to\infty} R(x_n) = f(2)$. Can we conclude that R is continuous at $x = 2$? Definitely not! The definition of continuity requires that the sequence of function values converges to the actual function value for *any* input sequence, and we have only tried one sequence.

Let's have another try without tying ourselves down to a particular input sequence such as $\{2 + 1/n\}$. Let $\{x_n\}$ be any sequence which converges to 2. Then it is reasonable to suppose that $\{x_n^2\}$ converges to 2^2 and hence that $\{4.9x_n^2\}$ converges to 4.9×2^2. Hence $\{R(x_n)\}$ converges to $R(2)$. Since $\{x_n\}$ can be any sequence converging to 2, we conclude that R is continuous at 2. □

In this last example we have made assumptions about the convergence of combinations of sequences. For example, we assumed $\{x_n^2\}$ converges to x_0^2 if $\{x_n\}$ converges to x_0. There is a theorem which deals with these matters. We state it without proof.

THEOREM 3.1

Let $\{x_n\}$ be a sequence converging to x_0 and let $\{y_n\}$ be a sequence converging to y_0. Then

(a) $\{kx_n\}$ converges to kx_0 for any constant k

(b) $\{x_n + y_n\}$ converges to $x_0 + y_0$

(c) $\{x_n - y_n\}$ converges to $x_0 - y_0$

(d) $\{x_n y_n\}$ converges to $x_0 y_0$

(e) $\{x_n/y_n\}$ converges to x_0/y_0 if $y_0 \neq 0$ and $y_n \neq 0$ for any n.

This theorem can be proved using a careful definition of convergence of sequences, but we shall not give the proof here. It is often possible to use this theorem to prove results on continuity, as we will show below.

EXAMPLE 3.4

In order to decide whether or not the sequence

$$\{a_n\} = \left\{\frac{n}{n+1}\right\}$$

converges, we note that

$$\frac{n}{n+1} = \frac{n+1-1}{n+1}$$
$$= 1 - \frac{1}{n+1}.$$

It should be clear (and this can be proved) that

$$\lim_{n\to\infty} \frac{1}{1+n} = 0,$$

so that
$$\lim_{n\to\infty} \frac{n}{1+n} = \lim_{n\to\infty}\left(1 - \frac{1}{1+n}\right)$$
$$= 1 - 0$$
$$= 1.$$

Hence the sequence $\{a_n\}$ converges to 1.

EXAMPLE 3.5
Suppose we want to investigate the behaviour of the function given by the rule $f(x) = x^2$ near the point $x = 3$. We may be interested in investigating properties of the function such as continuity or differentiability at the point $x = 3$. To this end, it is often useful to take a sequence $\{x_n\}$ converging to 3 and see what happens to the sequence $\{f(x_n)\}$.

First, note that $f(3) = 9$, so that the function is defined at 3. Next, we choose an input sequence that converges to 3 from above and see what happens to the corresponding sequence of function values. Let
$$\{x_n\} = \left\{3 + \frac{1}{n}\right\}.$$
Then $\{x_n\}$ is a sequence converging to 3 from above and we have
$$\{f(x_n)\} = \left\{\left(3 + \frac{1}{n}\right)^2\right\}$$
$$= \left\{9 + \frac{6}{n} + \frac{1}{n^2}\right\}$$
$$\to 9 \text{ as } n \to \infty.$$

We see that $\{f(x_n)\} \to f(3)$ as $n \to \infty$.

Next, we choose an input sequence that converges to 3 from below and see what happens to the corresponding sequence of function values. Let
$$\{t_n\} = \left\{3 - \frac{1}{n}\right\}.$$
Then $\{t_n\}$ is a sequence converging to 3 from above. We have
$$\{f(t_n)\} = \left\{\left(3 - \frac{1}{n}\right)^2\right\}$$
$$= \left\{9 - \frac{6}{n} + \frac{1}{n^2}\right\}$$
$$\to 9 \text{ as } n \to \infty.$$

We see that $\{f(t_n)\} \to f(3)$ as $n \to \infty$.

We have looked at two input sequences, which converge to 3. In both cases the corresponding function values converge to $f(3)$ and it seems plausible that f is continuous at $x = 3$, but we have not proved this. □

In the next example, we show how Theorem 3.1 may be used to prove continuity.

EXAMPLE 3.6
Let $f(x) = x^2$. Show that f is continuous at $x = 3$.
Solution. To show that a function f is continuous at a point x_0, we need to show that the output sequence $\{f(x_n)\}$ converges to $f(x_0)$ for every sequence $\{x_n\}$ which converges to x_0.

Let $x_0 = 3$ and choose *any* sequence $\{f(x_n)\}$ converging to 3. Then, by Theorem 3.1 (d)

$$\begin{aligned} \{f(x_n)\} &= \{x_n^2\} \\ &= \{x_n \times x_n\} \\ &\to 3 \times 3 \end{aligned}$$

so that $\{f(x_n)\} \to f(3)$ for any sequence $\{x_n\}$ converging to 3. The conclusion is that f is continuous at 3.

EXAMPLE 3.7
Show that

$$f(x) = 3x^2 + 5x$$

is continuous at any point x_0.
Solution. Let $\{x_n\}$ be a sequence converging to x_0. Then using Theorem 3.1, $\{5x_n\}$ converges to $5x_0$ by part (a), $\{x_n^2\}$ converges to x_0^2 by part (d), so by part (a) again, $\{3x_n^2\}$ converges to $3x_0^2$. Hence, by part (b), $\{3x_n^2 + 5x_n\}$ converges to $3x_0^2 + 5x_0$. Since this is true for any input sequence converging to x_0, it follows that $f(x) = 3x^2 + 5x$ is continuous at x_0.

EXAMPLE 3.8
Define a function f by

$$f(x) = \begin{cases} 0 & \text{if } x < 0, \\ x & \text{if } 0 \leqslant x < 1, \\ x - 1 & \text{if } 1 \leqslant x < 2, \\ 3 & \text{if } x \geqslant 2. \end{cases}$$

We shall show that this function is not continuous at $x = 1$. We do this by taking the input sequence $\{1 - 1/n\}$. The output sequence is also $\{1 - 1/n\}$, which converges to 1. However, the function value at $x = 1$ is $f(1) = 0$, so the output sequence does not converge to the function value and the function is not continuous at $x = 1$. Notice, however, that if we had taken the input sequence $\{1 + 1/n\}$, then the output sequence would have been $\{(1 + 1/n) - 1\} = \{1/n\}$, which converges to 0, so that for this input sequence, the output sequence does converge to the function value. This illustrates the fact that we cannot conclude that a function is continuous on the basis of just one input sequence (or even 1000 input sequences). We can, however, conclude that it is not continuous on the basis of just one input sequence. □

These examples illustrate the roles played by definitions and theorems. Definitions provide us with the concepts and state precisely what we mean by the concepts, but they are not always the best tools to use in practice. We can prove theorems using the definitions and it is these theorems which provide the practical tools. Note, however, that the use of definitions can be effective in demonstrating that a property (such as continuity) does *not* hold. This happened with Example 3.8.

EXAMPLE 3.9
The function
$$f(x) = \frac{x^2 - x - 2}{(x+2)(x-1)}$$
of Example 2.15 is not continuous at the points -2 and 1. This is because these points do not lie in the domain of the function. No sequence of output values can converge to $f(-2)$ and $f(1)$, since these numbers do not exist.

EXAMPLE 3.10
The function
$$g(x) = \frac{x^2 - 4}{x - 2}$$
is not continuous at $x = 2$, since this point is not in the natural domain of the function. However, when $x \neq 2$ we can write
$$\begin{aligned} g(x) &= \frac{x^2 - 4}{x - 2} \\ &= \frac{(x-2)(x+2)}{x-2} \\ &= x + 2, \end{aligned}$$
so that $g(x_n)$ converges to 4 for every sequence $\{x_n\}$ converging to 2. Hence if we define $g(2)$ to be 4, then g is continuous everywhere, including the point $x = 2$. □

It is not hard to see that every polynomial function is continuous at every point of its domain. Such functions illustrate the idea of continuity on an interval.

> **DEFINITION 3.4 Continuity on an open interval**
> *A function which is continuous at every point of an **open** interval I is said to be **continuous on I**.*

It is also possible to define continuity on *closed* intervals.

> **DEFINITION 3.5 Continuity on a closed interval**
> *A function f is continuous on the closed interval $[a, b]$ in its domain if:*
>
> *1. f is continuous on the open interval (a, b).*
>
> *2. For every sequence $\{x_n\}$ in $[a, b]$ converging to a, the sequence $\{f(x_n)\}$ converges to $f(a)$.*
>
> *3. For every sequence $\{x_n\}$ in $[a, b]$ converging to b, the sequence $\{f(x_n)\}$ converges to $f(b)$.*

It should be clear how to define continuity on intervals such as $[a, b)$ or $(a, b]$ and we leave this as an exercise. A function which is continuous on an interval I (closed, open or neither) is said to be *continuous on I*. A function which is continuous at all points of its domain is simply referred to as being a *continuous function*. Thus polynomials are continuous functions.

EXERCISES 3.2

In Exercises 1–6, the first few terms of a sequence are given. Find an expression for the general term of each sequence, assuming the pattern continues as indicated.

1. $1, 4, 9, \ldots$
2. $1, \frac{1}{2}, \frac{1}{3}, \frac{1}{4}, \ldots$
3. $\frac{1}{2}, \frac{1}{6}, \frac{1}{12}, \frac{1}{20}, \frac{1}{30}, \frac{1}{42}, \ldots$
4. $-\frac{1}{4}, \frac{2}{9}, -\frac{3}{16}, \frac{4}{25}, -\frac{5}{36}, \ldots$
5. $\frac{1}{2}, -\frac{2}{3}, \frac{3}{4}, -\frac{4}{5}, \ldots$
6. $1, 1, 1, 1, \ldots$

In Exercises 7–12, find the first four terms of the sequence whose general term is given.

7. $\dfrac{1}{2^n}$
8. $\dfrac{1}{(n+1)(n+3)}$
9. $\dfrac{(-1)^n(n+1)}{n}$
10. $1 - \dfrac{1}{3^n}$
11. $\dfrac{n + (-1)^{n+1}}{n}$
12. $\sqrt{n+1}$

In Exercises 13–20, decide whether the given sequence is convergent or divergent. If it is convergent, find its limit.

13. $\left\{ 1 + \dfrac{1}{n} \right\}$
14. $\left\{ \dfrac{1 + n^2}{n} \right\}$
15. $\left\{ \sqrt{n+1} - \sqrt{n} \right\}$
16. $\left\{ \dfrac{3n^2 + 4n + 1}{n^2 + 1} \right\}$
17. $\left\{ \sqrt{\dfrac{3n}{n+1}} \right\}$
18. $\left\{ \left(1 - \dfrac{1}{3^n}\right)\left(2 + \dfrac{1}{3^n}\right) \right\}$
19. $\left\{ \dfrac{2^n}{n^2} \right\}$
20. $\left\{ \dfrac{n!}{n^n} \right\}$

In Exercises 21–25, the natural domain of the given function does not include the point $x = 1$. In each case, where possible, define $f(1)$ so that f is continuous at $x = 1$.

21. $f(x) = \dfrac{x^2 - 1}{x - 1}$
22. $f(x) = \dfrac{x - 1}{|x - 1|}$
23. $f(x) = \dfrac{1 - x^3}{x - 1}$
24. $f(x) = \dfrac{x^2 + x - 2}{x - 1}$
25. $f(x) = \dfrac{\sqrt{x + 3} - 2}{x - 1}$

26. The parking area in the Sydney Fish Markets charges $2 for the first three hours or part thereof and $10 per hour after the third hour, up to a maximum of $24.

    (a) Sketch the graph of the cost of parking charges as a function of time.
    (b) At what points is this function discontinuous? What is the significance of these discontinuities to someone who uses the parking area?

27. Let f be the function defined by the rule

$$f(x) = \begin{cases} x^2, & x \leqslant 2 \\ x+1, & x > 2. \end{cases}$$

Draw the graph of f. It should be clear from your graph that f is not continuous at $x = 2$. To verify this you need to find a sequence $\{x_n\}$ which converges to 2, but for which $\{f(x_n)\}$ does not converge to $f(2) = 4$. Show that $\{2 + 1/n\}$ is such a sequence. What does $\{f(2+1/n)\}$ converge to? Find a sequence $\{x_n\}$ for which $\{f(x_n)\}$ converges to 4.

28. Two important functions in science and engineering are the *Heaviside function* H_a, defined by

$$H_a(t) = \begin{cases} 0, & t < a \\ 1, & t \geqslant a \end{cases}$$

for any $a \in \mathbb{R}$, and the *unit pulse function* $P_{\epsilon,a}$, defined by

$$P_{\epsilon,a}(t) = \tfrac{1}{\epsilon}[H_a(t) - H_{a+\epsilon}(t)]$$

for any $a, \epsilon \in \mathbb{R}$. Draw the graph of each function and determine where the functions are continuous.

29. Let H_a be the Heaviside function defined above. Draw the graphs of $H_2(t)$, $H_2(1+t)$ and $H_{-1}(-t)$.

30. Explain how we can define continuity of a function on intervals such as $[a, b)$ or $(a, b]$.

For each sequence $\{x_n\}$ of input numbers in Exercises 31–35, find the corresponding sequence of output numbers $\{f(x_n)\}$ for the functions indicated.

31. $\{x_n\} = \left\{5 + \dfrac{1}{n}\right\}$, $f(x) = 3 + x^2$

32. $\{x_n\} = \left\{5 - \dfrac{1}{n}\right\}$, $f(x) = 3 + x^2$

33. $\{x_n\} = \left\{1 - \dfrac{1}{n}\right\}$, $f(x) = x^3$

34. $\{x_n\}$, $f(x) = x^2$

35. Show that the function defined by $f(x) = 5 + x$ is continuous at $x = 7$.

36. Show that the function defined by $f(x) = 5 + x$ is continuous at $x = x_0$.

37. Show that the function defined by

$$f(x) = \begin{cases} x^2, & x \leqslant 2 \\ 7 - x, & x > 2 \end{cases}$$

is not continuous at $x = 2$.

38. Show that the function

$$f(x) = \begin{cases} x & \text{if } x < 1 \\ x - 1 & \text{if } x \geqslant 1 \end{cases}$$

is continuous on each of the intervals $(-\infty, 1)$ and $[1, \infty)$, but not on the interval $(0, 2)$.

CHAPTER 4
DIFFERENTIATION

In Chapter 2, we informally discussed continuity and differentiability. In Chapter 3, we introduced precise definitions for continuity and we must now do the same for differentiability. In Chapter 2, we informally defined a function to be differentiable at a point if its graph has no corner at that point. In this chapter, we present a more detailed discussion of differentiability.

4.1 THE DERIVATIVE

Let f be a given continuous function and let x_0 be a point in its domain. To investigate the differentiability at the point x_0, first take a point $x_n \neq x_0$ in the domain of f (Figure 4.1). The gradient of the secant ABC joining the points $(x_0, f(x_0))$ and $(x_n, f(x_n))$ is given by

$$m = \frac{f(x_n) - f(x_0)}{x_n - x_0}.$$

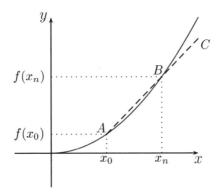

Figure 4.1: A secant of a graph

If we take an input sequence $\{x_n\}$ which converges to x_0 then for each point x_n we can take a secant joining $(x_0, f(x_0))$ and $(x_n, f(x_n))$. The gradients of such secants form a sequence

$$\left\{ \frac{f(x_n) - f(x_0)}{x_n - x_0} \right\}. \tag{4.1}$$

There are some minor sources of difficulty which we need to dispose of. In the first instance, we note that it is quite possible to choose a sequence $\{x_n\}$ which converges to x_0 and for which $x_m = x_0$ for at some positive integer m, in which case the corresponding quotient in the sequence (4.1) will be undefined. To avoid this happening, we shall restrict ourselves to sequences $\{x_n\}$ which have the properties $x_n \to x_0$ as $n \to \infty$ and $x_n \neq x_0$ for all positive integers n. Such sequences will be called *special* sequences converging to x_0.

In order for there to be sequences $\{x_n\} \to x_0$ with $x_n \neq x_0$, we shall require that f be defined on an open interval about x_0.

If x_0 is in the domain of a function f, and if $\{x_n\}$ is any special sequence converging to x_0, then several different things may happen:

- The sequence (4.1) may diverge for some (perhaps every) choice of $\{x_n\}$.

- The sequence (4.1) may converge to a number which depends on the choice of $\{x_n\}$.

- The sequence (4.1) may converge to the same number for every possible choice of $\{x_n\}$.

From a graphical point of view, it appears that if the sequence (4.1) of gradients of secants converges, then it will converge to the gradient of the tangent at x_0.

EXAMPLE 4.1
In the table, we have computed the sequence (4.1) for $f(x) = x^3$, $x_0 = 2$ and $\{x_n\} = \{2 + 1/n\}$.

n	1	2	5	10	100	1000
x_n	3	2.5	2.2	2.1	2.01	2.001
$(f(x_n) - f(2))/(x_n - 2)$	19	15.25	13.24	12.61	12.0661	12.0067

This table suggests (but does not prove) that

$$\lim_{n \to \infty} \frac{f(x_n) - f(2)}{x_n - 2} = 12. \qquad \square$$

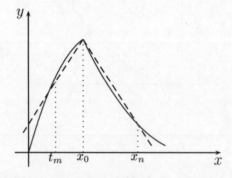

Figure 4.2: A nonsmooth graph

EXAMPLE 4.2
Consider the function whose graph is given in Figure 4.2, which has a corner at x_0. It is interesting to see how this corner manifests itself in terms of the gradients of secants. Firstly, if we take a sequence $\{x_n\}$ for which each $x_n > x_0$ and which converges to x_0, then the diagram suggests that the gradients of the corresponding secants will converge to a negative number. On the other hand, if we take a sequence $\{t_n\}$ with each $t_n < x_0$ and which converges to x_0, then the diagram suggests that the gradients of the corresponding secants will converge to a positive number. Clearly, the two gradients will not then converge to the same number.

EXAMPLE 4.3
Let us define the function f by the rule $f(x) = x^2$. This is a function whose graph we would certainly expect to satisfy our intuitive notion of a smooth function. In Example 4.1 above, we estimated the gradient at a point on the graph of $y = x^3$ by drawing up a table, but such a table cannot give an exact answer for the gradient at a particular point. In this example, we will use what we know about convergence of sequences to get an exact answer for the gradient of the tangent at a particular point on the graph.

Let x_0 be any real number (so that x_0 is in the domain of f). Choose any input sequence $\{x_n\}$ converging to x_0. We then have

$$\left\{\frac{f(x_n) - f(x_0)}{x_n - x_0}\right\} = \left\{\frac{x_n^2 - x_0^2}{x_n - x_0}\right\}$$

$$= \left\{\frac{(x_n + x_0)(x_n - x_0)}{x_n - x_0}\right\}$$

$$= \{x_n + x_0\}$$

and this sequence clearly converges to $2x_0$ independently of the choice of the sequence $\{x_n\}$. Notice that $x_n - x_0$ can be cancelled, because for every n, $x_n - x_0 \neq 0$.

EXAMPLE 4.4
Here is an example of a function which we do not expect to be smooth. Let $f(x) = |x|$, that is, let

$$f(x) = \begin{cases} -x, & x < 0 \\ x & x \geqslant 0 \end{cases}$$

The graph has a corner at $x = 0$, so let us investigate the sequence

$$\left\{\frac{f(x_n) - f(x_0)}{x_n - x_0}\right\} \tag{4.2}$$

for the case $x_0 = 0$. First take a decreasing input sequence $\{x_n\} = \{1/n\}$. This input sequence converges to 0. The output sequence (4.2) corresponds in this case to

$$\left\{\frac{f(x_n) - f(0)}{x_n - 0}\right\} = \left\{\frac{f(1/n) - f(0)}{1/n - 0}\right\}$$

$$= \left\{\frac{1/n - 0}{1/n - 0}\right\},$$

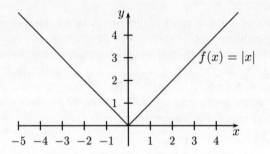

Figure 4.3: The graph of $f(x) = |x|$

which converges to 1. On the other hand, the increasing input sequence $\{y_n\} = \{-1/n\}$ also converges to 0, but the corresponding output sequence is

$$\left\{\frac{f(y_n) - f(0)}{y_n - 0}\right\} = \left\{\frac{f(-1/n) - f(0)}{-1/n - 0}\right\}$$
$$= \left\{\frac{1/n - 0}{-1/n - 0}\right\},$$

which converges to -1. In this example the presence of the corner has caused different output sequences to converge to different numbers, even though the corresponding input sequences both converged to $x = 0$. □

> **DEFINITION 4.1 Differentiable functions**
> Let f be a function which is defined at all points of an open interval I. The function f is **differentiable at $x_0 \in I$** if the sequence
> $$\left\{\frac{f(x_n) - f(x_0)}{x_n - x_0}\right\} \qquad (*)$$
> converges to the same limit for **every** special sequence $\{x_n\}$ converging to x_0, where $x_n \in I$ for all n. This limit is normally denoted by the symbol $f'(x_0)$.

Notice that the statement that the sequence

$$\left\{\frac{f(x_n) - f(x_0)}{x_n - x_0}\right\}$$

converges to $f'(x_0)$ can be written as

$$\lim_{n \to \infty} \frac{f(x_n) - f(x_0)}{x_n - x_0} = f'(x_0).$$

This definition says nothing about the interpretation of the value of the limit. From a graphical point of view, as x_n gets close to x_0, the terms of the sequence (*) in Definition 4.1 get close to the gradient of the tangent at x_0 for a function with no corner. We expect the limit of the sequence to be the slope of the tangent at x_0. A function with no corner at x_0 has a unique tangent, while if there is a corner, then there is not a unique tangent.

DEFINITION 4.2 The derivative
*The number $f'(x_0)$ is called the **derivative of the function f at the point** x_0. This enables us to define a new function (denoted by f') called the **derivative of** f according to the rule $x \mapsto f'(x)$, where x is any point in the domain of f at which f is differentiable. The domain of f' is the set of all points where f is differentiable.*

Notice that since f' is a function, it may also be differentiable. If this is the case, we denote its derivative by f'' and call it the *second derivative of f*. In the same way we can define third, fourth and higher derivatives.

EXAMPLE 4.5
Suppose $f(x) = x$. Let $\{x_n\}$ be a sequence converging to a point $x_0 \in \mathbb{R}$. Then
$$\frac{f(x_n) - f(x_0)}{x_n - x_0} = \frac{x_n - x_0}{x_n - x_0} = 1,$$
so that the sequence
$$\left\{ \frac{f(x_n) - f(x_0)}{x_n - x_0} \right\}$$
is just the sequence $\{1, 1, 1, \ldots\}$ which converges to 1. Thus $f'(x_0) = 1$.

EXAMPLE 4.6
If $f(x) = x^2$, then according to the calculations in Example 4.3, f is differentiable at all points of its domain and f' is the function with domain $\mathbb{R}$ given by the rule $f'(x) = 2x$.

EXAMPLE 4.7
If $f(x) = |x|$, then f is differentiable at all points of its domain, except for the point $x = 0$, so that f' has domain $\mathbb{R} - \{0\}$ and
$$f'(x) = \begin{cases} -1 & \text{if } x < 0 \\ 1 & \text{if } x > 0. \end{cases} \tag{4.3}$$
We leave you to prove this result as an exercise. □

The derivative of a differentiable function is one of its most important properties. If we are given some function, then the derivative can often tell us a great deal about the behaviour of the function and may assist in sketching its graph. There is a more important aspect to the derivative, however. In many problems we find that the natural formulation of the problem gives us information about the derivative of the solution function rather than information about the function itself; we have to use the derivative to calculate the solution function. This method of solving problems by first finding the derivative and then using this derivative to find the solution function is the basis of calculus. It was discovered about 300 years ago by Sir Isaac Newton and Gottfried Leibniz.

The definition we have given for the derivative of a function expresses precisely what we mean by a function having no corners in its graph. We can use this definition to calculate the derivative in a few simple cases, but it is not a very efficient method compared with those which we shall shortly develop. Using the definition to find derivatives is known as *differentiation by first principles*. In this book, we use differentiation by first principles in the following two instances:

- To aid in the understanding of the definition

- To prove theorems which enable us to use more efficient methods of calculating derivatives.

A process based on differentiation by first principles also occurs in some numerical work. When measuring a variable quantity such as a seismic signal, we may only have its values at discrete points in time. To calculate the time rate of variation, we have to approximate the derivative by the slope of a chord joining two nearby data points. This is the basis of *finite difference methods*.

4.1.1 Remarks on notation

Before going on to develop efficient rules for differentiation, we need to consider some matters of notation. All of the functions we have discussed have been denoted by a single letter such as f, h or ϕ. We can regard such a letter as the *name* of the given function. Of course, we don't have to restrict ourselves to single letters for function names. We could, for instance, define a function 'step' by the rule

$$\text{step}(t) = \begin{cases} 0, & t \leqslant 0 \\ 1, & t > 0. \end{cases}$$

Some functions have names which are universally agreed upon. You have probably met some of these before. Common examples are $\sin(x)$, $\cos(x)$ and $\log(x)$, usually written as $\sin x$, $\cos x$ and $\log x$. We don't even have to use letters, for example the square root function is denoted by the special symbol $\sqrt{}$ and the output number corresponding to the input number x is denoted by $\sqrt{x}$.

Many simple functions do not have universally agreed names and are defined by algebraic formulas, for example polynomials such as

$$f(x) = x^3 - x^2 + 2x - 1.$$

Indeed, we often only refer to the formula in speaking of the function. Thus we (incorrectly) speak of the function $x^3 - x^2 + 2x - 1$, when we really mean the function whose output number is $x^3 - x^2 + 2x - 1$ when the input number is x. Function values in such cases are denoted by a vertical stroke notation:

$$(x^3 - x^2 + 2x - 1)|_{x=10} = 919.$$

This means the same thing as $f(10)$ if $f(x) = x^3 - x^2 + 2x - 1$. *Mathematica* uses a similar notation. If we want to substitute $x = 10$ in the expression $x^3 - x^2 + 2x - 1$, then the *Mathematica* expression is

```
x^3-x^2+2x-1/.x->10
```

Another notation is often used in applications where there are independent and dependent variables. If the values of the independent variable are denoted by x and those of the dependent variable by y and if the function is defined by a formula such as x^2, then we write $y = x^2$ and say that y is a function of x. Particular values of y are denoted by the function value notation. Thus $y(3) = 3^2 = 9$. The derivative is written y' and this again denotes a function with particular values written with the function value notation, as in $y'(2)$.

There is also an alternative notation for derivatives which originated with Leibniz and which is still used today—the dy/dx notation. Recall that the derivative of a function f at a point x_0 is

$$f'(x_0) = \lim_{n \to \infty} \frac{f(x_n) - f(x_0)}{x_n - x_0},$$

where $\{x_n\}$ is any special sequence that converges to x_0. Let us put $y = f(x)$. Leibniz wrote $\Delta x = x_n - x_0$ and $\Delta y = f(x_n) - f(x_0)$ and considered the quotient $\Delta y / \Delta x$. As $n \to \infty$ this quotient approaches $f'(x_0)$, but instead of writing $f'(x_0)$ for the derivative, Leibniz wrote $\dfrac{dy}{dx}$ and called it the *derivative of y with respect to x*. Note that $\dfrac{dy}{dx}$ does not mean dy divided by dx. It is just another way of writing f', where there is no suggestion of division. In the Leibniz notation, the derivative of the function

$$y = x^3 - x^2 + 2x - 1$$

would be written as

$$\frac{dy}{dx}$$

or

$$\frac{d}{dx}(x^3 - x^2 + 2x - 1)$$

and this denotes the same function as f' if we had put $f(x) = x^3 - x^2 + 2x - 1$. Function values of the derivative are written with a vertical stroke. Thus

$$\left.\frac{dy}{dx}\right|_{x=2}$$

or

$$\left.\frac{d}{dx}(x^3 - x^2 + 2x - 1)\right|_{x=2}$$

is the same thing as $f'(2)$. As simple examples we have

$$x^2|_{x=3} = 9,$$

$$\frac{d}{dx}x^2\Big|_{x=3} = 2x|_{x=3} = 6.$$

EXERCISES 4.1

In Exercises 1–6, find $f'(3)$ by using the definition of a derivative.

1. $f(x) = 4x + 1$
2. $x^3 + x$
3. $f(x) = \sqrt{x} + 1$
4. $f(x) = 1/(x+2)$
5. $f(x) = (2x+3)^2$
6. $1/x$

In Exercises 7–12, differentiate the given function by using the definition of a derivative. In each case, plot the graph of the derivative.

7. $f(x) = 3$
8. $f(x) = x + 1$
9. $f(x) = 3x - 2$
10. $f(x) = 1/(x+2)$
11. $g(x) = 2x^2 + 1$
12. $h(x) = x^3 + 4x + 1$
13. Show that if $f(x) = |x|$, then

$$f'(x) = \begin{cases} -1 & \text{if } x < 0 \\ 1 & \text{if } x > 0. \end{cases}$$

Plot the graph of f'. At what points is f' differentiable?

14. Decide which, if either, of the two functions $g(x) = |x^2|$ and $h(x) = |x^3|$ is differentiable at $x = 0$; justify your answers.

In Exercises 15–18, draw the graph of the derivative of the given function.

15.
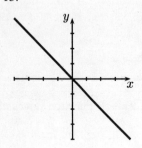

16.
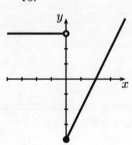

17.
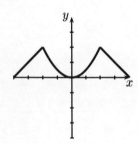

18.
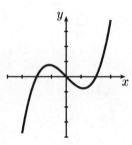

19. Let f be a differentiable function and let h be a small positive number. Give a physical meaning to the quotient
$$\frac{f(x+h) - f(x)}{h}$$
in the following cases.

   (a) $f(x)$ is the height of a missile x seconds after liftoff.

   (b) $f(x)$ is the population of a city at time x.

   (c) $f(x)$ is the cost in dollars of producing x units of a product.

4.2 RULES FOR DIFFERENTIATION

We have identified the derivative of a differentiable function as one of its important properties. As yet, we have no way of calculating the derivative, except by the use of Definition 4.1 on page 44, which is often a cumbersome process. In this section we will develop more efficient computational procedures for evaluating derivatives.

The functions we have already met have often been combinations of simpler functions. For instance, the function
$$f(x) = (x+1)^3 + (x+2)^2$$

is the sum of the two functions $f_1(x) = (x+1)^3$ and $f_2(x) = (x+2)^2$, while the function
$$F(x) = (x+1)^3(x+2)^2$$
is the product of f_1 and f_2. It turns out that we can use the derivatives of f_1 and f_2 to find the derivatives of f and F. This means that with a relatively small number of derivatives and rules for the differentiation of combinations of functions, we can differentiate quite complicated functions.

> **THEOREM 4.1 The derivative of a constant function**
> Suppose f is a constant function given by $f(x) = k$ for all x and some real number k. Then $f'(x) = 0$ for all x.

The proof is left as an exercise.

One of the most common ways of forming new functions is by using sums, products and quotients of given functions, so let us now be precise about such combinations of functions and their derivatives.

> **DEFINITION 4.3 Sums of functions**
> Let f and g be functions. The sum of f and g is another function, denoted by $f + g$, defined by the rule $(f+g)(x) = f(x) + g(x)$. The domain of $f + g$ is the set of all points where f and g are both defined.

EXAMPLE 4.8
Let $f(x) = x^2 + 2x$ and let $g(x) = x^3 + 2x^2 + 1$. Then
$$\begin{aligned}(f+g)(x) &= f(x) + g(x) \\ &= (x^2 + 2x) + (x^3 + 2x^2 + 1) \\ &= x^3 + 3x^2 + 2x + 1.\end{aligned}$$

> **THEOREM 4.2 The derivative of a sum**
> Suppose f and g are both differentiable at a point x. Then $f + g$ is also differentiable at x and $(f+g)'(x) = f'(x) + g'(x)$.

We prove this result by using Definition 4.1 on page 44 and Theorem 3.1 on page 34. Let $\{x_n\}$ be a sequence converging to x. Then
$$\begin{aligned}\frac{(f+g)(x_n) - (f+g)(x)}{x_n - x} &= \frac{f(x_n) + g(x_n) - (f(x) + g(x))}{x_n - x} \\ &= \frac{f(x_n) - f(x)}{x_n - x} + \frac{g(x_n) - g(x)}{x_n - x}.\end{aligned}$$

The first term on the right hand side approaches $f'(x)$ as $n \to \infty$, while the second approaches $g'(x)$. Thus $(f+g)'(x) = f'(x) + g'(x)$.

An alternative and traditional way of writing the above result is to put $u = f(x)$ and $v = g(x)$. Then
$$\frac{d}{dx}(u+v) = \frac{du}{dx} + \frac{dv}{dx}.$$

EXAMPLE 4.9
As a very simple example we have
$$\frac{d}{dx}(x^2 + x) = 2x + 1. \qquad \square$$

We can also define *scalar multiples* of functions: if k is any real number and f is any given function, then kf is the function defined by

$$(kf)(x) = kf(x).$$

It has the same domain as f. For example if $f(x) = x^2$ then $4f$ is the function given by the rule $x \mapsto 4x^2$. It is easy to verify that, for any real number k,

$$(kf)'(x) = kf'(x)$$

and you should do this as an exercise.

We can summarise the two rules we have so far:

- The derivative of a sum is equal to the sum of the derivatives.

- The derivative of a scalar multiple of a function is the same scalar multiple of the derivative of the function.

Unfortunately, the results for products and quotients do not follow such a simple form. First a definition:

DEFINITION 4.4 Products and quotients of functions
Let f and g be two functions.

- *The **product** of f and g is the function fg defined by the rule $(fg)(x) = f(x)g(x)$. The product is defined at all points x where f and g are both defined.*

- *The **quotient** of f and g is the function $\dfrac{f}{g}$ defined by the rule*

$$\left(\frac{f}{g}\right)(x) = \frac{f(x)}{g(x)}.$$

The quotient is defined for all points x where f and g are both defined and where $g(x) \neq 0$.

EXAMPLE 4.10
Let $f(x) = x^2$ and let $g(x) = x^2 - 1$. Then

$$(fg)(x) = x^2(x^2 - 1).$$

The domain of both f and g is $\mathbb{R}$, so that the domain of fg is also $\mathbb{R}$.

As for the quotient of these functions, we have

$$\left(\frac{f}{g}\right)(x) = \frac{x^2}{x^2 - 1}.$$

The domain of the quotient function is $\mathbb{R} \setminus \{-1, 1\}$. □

THEOREM 4.3 The derivative of a product
Suppose f and g are both differentiable at a point x. Then the product fg is also differentiable at x and $(fg)'(x) = f'(x)g(x) + f(x)g'(x)$.

To justify this result, we take a point x in the domain of fg and an input sequence $\{x_n\}$ converging to x. Then we must consider the sequence whose general term is

$$\frac{(fg)(x_n) - (fg)(x)}{x_n - x} = \frac{f(x_n)g(x_n) - f(x)g(x)}{x_n - x}. \tag{4.4}$$

The aim of the game is to write the left hand side of equation (4.4) in terms of

$$\frac{f(x_n) - f(x)}{x_n - x}$$

and

$$\frac{g(x_n) - g(x)}{x_n - x},$$

since these terms converge to $f'(x)$ and $g'(x)$ respectively. There is a trick used here: subtract and then add the term $f(x)g(x_n)$ in the numerator of equation (4.4). This gives

$$\frac{(fg)(x_n) - (fg)(x)}{x_n - x} = \frac{f(x_n)g(x_n) - f(x)g(x_n) + f(x)g(x_n) - f(x)g(x)}{x_n - x}$$
$$= \frac{f(x_n) - f(x)}{x_n - x} g(x_n) + \frac{g(x_n) - g(x)}{x_n - x} f(x). \tag{4.5}$$

Now as $n \to \infty$,

$$\frac{f(x_n) - f(x)}{x_n - x} \to f'(x)$$

and

$$\frac{g(x_n) - g(x)}{x_n - x} \to g'(x),$$

but what does $g(x_n)$ approach as $n \to \infty$? We know that if g is continuous at x, then $g(x_n)$ approaches $g(x)$ as x_n approaches x, but how can we assume that g is continuous at x? Well, it is true in general that if a function is differentiable at a point x then it is automatically continuous there. In fact

$$g(x_n) = \frac{g(x_n) - g(x)}{x_n - x}(x_n - x) + g(x)$$
$$\to g'(x) \times 0 + g(x)$$
$$= g(x),$$

where we have used Theorem 3.1. Thus the right hand side of equation (4.5) approaches $f'(x)g(x) + f(x)g'(x)$ and this is the derivative of $(fg)(x)$. □

Theorem 4.3 can be written in Leibniz notation as

$$\frac{d}{dx}(uv) = u\frac{dv}{dx} + v\frac{du}{dx},$$

where $u = f(x)$ and $v = g(x)$. Note that the derivative of the product is not the product of the derivatives. The correct rule, as given by Theorem 4.3 is an inevitable consequence of the definition of the derivative and its interpretation as the slope of the tangent or (as we shall see in the next section) as a rate of change.

EXAMPLE 4.11
We have

$$\frac{d}{dx}(x^2) = \frac{d}{dx}(x \times x)$$
$$= x\frac{d}{dx}x + x\frac{d}{dx}x$$
$$= x \times 1 + x \times 1 = 2x,$$

the same result we obtained earlier by first principles. Next we have

$$\frac{d}{dx}(x^3) = \frac{d}{dx}(x^2 \times x)$$
$$= x^2\frac{d}{dx}x + x\frac{d}{dx}x^2$$
$$= x^2 \times 1 + x \times 2x = 3x^2,$$

and in general

$$\frac{d}{dx}(x^n) = nx^{n-1}.$$ □

The last result in the above example is an important one and is worth highlighting.

THEOREM 4.4
If n is a positive integer and $f(x) = x^n$ for $x \in \mathbb{R}$, then $f'(x) = nx^{n-1}$.

There are a number of ways of proving this result. One way has been suggested in the above example and another method is outlined in Exercise 4.2.

Finally, we state the rule for differentiating a quotient. This rule can be proved along the same lines as for a product, but we shall not present the proof here.

THEOREM 4.5 The derivative of a quotient
Suppose f and g are both differentiable at a point x and suppose $g(x) \neq 0$. Then the quotient f/g is also differentiable at x and

$$\left(\frac{f}{g}\right)'(x) = \frac{f'(x)g(x) - f(x)g'(x)}{g(x)^2}.$$

This result can be written in the Leibniz notation as

$$\frac{d}{dx}\left(\frac{u}{v}\right) = \frac{vu' - uv'}{v^2},$$

where $u = f(x)$ and $v = g(x)$.

With the results in this section, we can differentiate any function defined by an algebraic formula without having to go back to the original definition. We can do this because we have proved some theorems. The purpose of theorems is to simplify calculations. We had to use the original definition in the proofs of the theorems, but once we have proved the theorems we no longer have to differentiate by first principles. Of course, if we encounter a new function which is not simply a combination of known ones, then we would have to go back to the definition to find its derivative.

EXAMPLE 4.12
Let
$$f(x) = \frac{x^2}{x^2 - 1}.$$
Then
$$\begin{aligned} f'(x) &= \frac{(x^2-1)\frac{d}{dx}(x^2) - x^2\frac{d}{dx}(x^2-1)}{(x^2-1)^2} \\ &= \frac{2x(x^2-1) - (2x)(x^2)}{(x^2-1)^2} \\ &= -\frac{2x}{(x^2-1)^2}. \end{aligned}$$

EXERCISES 4.2

In the following Exercises, differentiate the given function using the theorems in Section 4.2.

1. $f(x) = x^2 + 3x$
2. $f(x) = 2x^3 + 3x^2 + 1$
3. $f(x) = \dfrac{1}{1+x}$
4. $f(x) = (2x+1)(x^2-1)$
5. $f(x) = \dfrac{x^2+2}{x-1}$
6. $f(x) = (4x^5 + x)(2x+1)$
7. Show that if $f(x) = k$ for all x, where k is some constant, then $f'(x) = 0$.
8. Here is an outline of a proof that if $f(x) = x^n$, then $f'(x) = nx^{n-1}$, where n is a positive integer. Begin by checking that
$$b^n - a^n = (b-a)(b^{n-1} + ab^{n-2} + a^2b^{n-3} + \cdots$$
$$\cdots + a^{n-3}b^2 + a^{n-2}b + a^{n-1})$$
for any real numbers a and b and any positive integer n. Next, let $\{x_k\}$ be an arbitrary sequence which converges to x. Show that
$$\frac{f(x_k) - f(x)}{x_k - x} = x_k^{n-1} + xx_k^{n-2} + \cdots + x^{n-3}x_k^2 + x^{n-2}x_k + x^{n-1},$$
and hence that the sequence
$$\left\{ \frac{f(x_k) - f(x)}{x_k - x} \right\}$$
converges to nx^{n-1}.
9. Use Theorem 4.5 to show that if $f(x) = x^n$, then $f'(x) = nx^{n-1}$, where n is a *negative* integer.

4.3 VELOCITY, ACCELERATION AND RATES OF CHANGE

Historically, the derivative first explicitly appeared in connection with the motion of bodies. This was an innovation by Sir Isaac Newton who, in the period 1665–1667, laid the foundations of the calculus.

Newton was interested in the mechanics of motion, particularly the motion of the planets around the sun, which involved velocity and acceleration. He restricted his attention to derivatives with respect to time, which he called *fluxions*. His notation of $\dot{x}$ for dx/dt is still widely used today.

4.3.1 VELOCITY

Suppose a particle moves along a straight line in such a way that its distance s from a fixed point O at time t is given by $s = f(t)$, where f is a smooth function. In general the velocity at which the particle moves will vary as the time changes and we may wish to find the velocity at a particular time t_0, say. We can do this in the following way. First take a special sequence $\{t_n\}$ which converges to t_0. For any particular value of n, the *average* velocity between the time t_0 and the time t_n is given as

$$\text{average velocity} = \frac{\text{distance moved}}{\text{time taken}}$$
$$= \frac{f(t_n) - f(t_0)}{t_n - t_0}.$$

As we let n get arbitrarily large, the time interval $t_n - t_0$ will decrease and we expect that the sequence of average velocities will converge. The number to which it converges is defined to be the instantaneous velocity at the time t_0. But the number to which the sequence

$$\left\{ \frac{f(t_n) - f(t_0)}{t_n - t_0} \right\}$$

converges is just $f'(t_0)$, so that at the time t_0, the velocity v is given by

$$v = f'(t_0).$$

4.3.2 RATES OF CHANGE

In the case of a straight line $y = mx + c$, the slope m of the line gives the rate of change of y with respect to x. If x changes by an amount Δx then the corresponding change Δy in y is m times as much, that is, $\Delta y = m\Delta x$. The *sign* of m is significant. If it is negative, then y decreases as x increases, while if it is positive, then y increases as x increases.

In the more general case of a differentiable function $y = f(x)$, we can proceed in a similar way to which we did in the case of velocities. We define the *instantaneous rate of change* of y with respect to x at the point x_0 by $f'(x_0)$. If y is increasing at x_0, then the rate of change is positive, while if y is decreasing at x_0, the rate of change is negative. As an example, velocity is the rate of change of distance with respect to time.

EXAMPLE 4.13

A screen saver consists of expanding squares which appear at random points on a computer screen. If the length of side x of each square increases in length with a speed of 2 cm/sec, find the rate of change of the area of a given square when its side length is 6 cm.
Solution. The area A of a given square is given by

$$A = x^2$$
$$= 4t^2,$$

since $x = 2t$ at time t. Consequently,

$$A'(t) = 8t.$$

Since $t = 3$ when $x = 6$, we compute $A'(3) = 24$. Hence the rate of change of area with respect to time is $24\,\text{cm}^2/\text{sec}$.

4.3.3 ACCELERATION

Finally we consider the concept of *acceleration*. If a particle has a velocity v_1 at time t_1 and a velocity v_2 at time t_2, then the average acceleration during the time interval $t_2 - t_1$ is defined to be

$$\frac{v_2 - v_1}{t_2 - t_1}.$$

By using a similar argument to the above, we define the instantaneous acceleration a at a time t_0 to be

$$a = \left.\frac{dv}{dt}\right|_{t_0} = f''(t_0).$$

The acceleration is the rate of change of velocity with respect to time.

EXERCISES 4.3

Exercises 1–4 refer to a body which moves along the x axis in such a way that its position at time t is given by $x = t^2 - t$.

1. Find the average velocity of the body over the time interval $[t_0, t_0 + h]$.
2. Tabulate the average velocities over the time intervals $[3, 3+h]$, for $h = 1, 0.1, 0.01$ and 0.001 seconds.
3. Use the results from Exercise 2 to guess the instantaneous velocity at $t = 3$.
4. Let $\{h_n\}$ be any sequence converging to zero. Confirm your guess in Exercise 3 by finding the limit of the average velocities over the time intervals $[3, 3 + h_n]$, $n = 1, 2, 3, \ldots, \infty$.
5. If an object has fallen a distance of $4.9t^2$ metres after t seconds, what is its velocity after 4 seconds?
6. The distance s of a particle from a fixed point at time t is given by

$$s = t^3 - 9t^2 + 15t.$$

   At what times is the velocity zero? What is the acceleration at these times?
7. Find the rate of change of the area of a circle with respect to the radius. What is this rate of change when the radius is $3\,\text{m}$?
8. Find the rate of change of the area of a square with respect to the length x of one of its diagonals. What is the rate of change when $x = 3$?
9. A pebble is dropped into a pond at time $t = 0$, where t is measured in seconds. This causes a circular ripple that moves out from the point of impact at $2\,\text{m/s}$. At what rate (in m^2/s) is the area within the circle increasing when $t = 5$?
10. Show that the rate of change of area of a circle with respect to its radius is equal to its circumference.
11. Show that the rate of change of volume of a sphere with respect to its radius is equal to its surface area.

12. Show that the rate of change of volume of a cube with respect to the length of its edge is equal to half its surface area.

13. The distance s (in metres) of a car from a fixed point at time t (in seconds) is given in the following table.

t	0	1	2	3	4	5	6	7	8	9	10	11	12
s	0	1.2	4.7	10.6	19.3	30.0	43.2	58.6	75.6	97.2	118.8	140.4	162.0

    Find the average velocity during the time interval $3 \leqslant t \leqslant 5$. Estimate the velocity and acceleration at $t = 6$ s. Over which period of time is the car moving with constant velocity?

14. The elevation of the Parramatta River s km from its source near Parramatta is given by $f(s)$ for some function f. What is a reasonable domain for f? What can you say about the sign of $f'(s)$?

15. The following graph shows the percentage of alcohol $A(t)$ in the blood of a person as a function of time t in hours since drinking two double whiskies. What was the approximate rate of increase in the percentage of alcohol in the blood after half an hour? At approximately what rate was the person eliminating alcohol after four hours? The legal blood limit in most states in Australia is 0.05%. When could this person legally drive home? [From Petocz, Petocz and Wood, *Introductory Mathematics*, Thomas Nelson, Melbourne, 1992.]

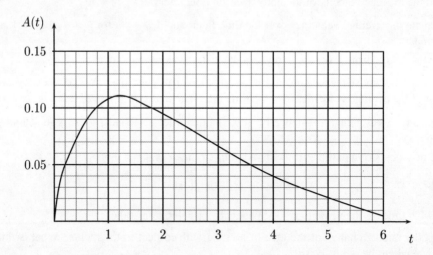

16. Newton's law of gravitation states that when a spaceship is at a distance r from the centre of the earth, its weight W is given by
$$W = \frac{GMm}{r^2},$$
where M is the mass of the earth, m is the mass of the spaceship and G is a constant, known as the gravitational constant. Find the rate of change of W with respect to r and show that it is negative. What is the physical meaning of this result?

CHAPTER 5
FALLING BODIES

We now have enough mathematical machinery available to solve the first of the problems posed in Chapter 1, that of falling bodies. In this chapter, we examine the motion of bodies moving under the influence of gravity. Here are a few of the cases where the mathematics needed for this type of problem is used:

- In the *Tower of Terror* described in Chapter 1
- In the motion of rockets and missiles
- In designing artillery
- In some movie stunts.

We have already set down the requirement that the solution to any of the problems we will consider in this book should be a smooth function, that is, a function which is continuous and differentiable. So far the only functions we have considered are those defined by algebraic formulas. There are three main types:

- Power functions of the form $f(x) = x^n$, where n is an integer. For example $f(x) = x^2$ or $f(x) = x^{-4}$.
- Polynomials, which are the sums of positive integer power functions. For example $f(x) = 2x^2 + 3x + 1$ or $f(x) = x^3 - 5$.
- Rational functions, which are quotients of polynomials. For example $f(x) = \dfrac{x^2 - x + 1}{x^2 + 3x + 2}$.

Such functions are differentiable at all points of their domain, which in the case of a rational function is the set of all points for which the denominator is not zero. Thus in searching for a solution to a practical problem, the first place we might look is among functions of the above types.

5.1 THE *TOWER OF TERROR*

Suppose, as in the *Tower of Terror*, we project a body upwards from some point on or above the Earth's surface and observe it rise, stop at its highest point and then fall. Let h be the height above the ground and let t be the time. We are seeking a function f such that $h = f(t)$. The rule defining f will tell us how to calculate values of h from values of t. Notice the role played by the various symbols: h is a *number* which is calculated from another *number* t by the application of a *rule* f. This is what we have to do if we wish to keep things strictly correct. However, what is correct and what is convenient may sometimes be two different things. In what follows, it will be more convenient to write $h(t)$ instead of $f(t)$. The symbol h is thus doing double duty. It may denote either a number or the rule for calculating this number. Mathematicians have introduced special terminology to describe such cases. They call it *abuse of notation*, as in "by abuse of notation, let us write $h(t)$ for $f(t)$".

The problem we are considering is a very old one. Throughout recorded history, people have been hurling objects through the air, usually at each other. It was regarded as important to understand matters such as the flight of arrows or cannonballs. The theory is now well understood, but this was not always the case. There were many misconceptions about the laws governing the motion of projectiles and it was not until the 17th century that Galileo finally developed the theory that we use today. Since we now know how to predict the path of projectiles we could simply tell you the relevant functions. However, this is only one of the many practical problems to which mathematics can be applied. By looking at the development of a simple problem such as this one, you will learn methods that will be applicable to more complex problems. We shall follow an idealised development of the problem and omit the errors and blind alleys that actually occurred in the development of the theory. Our aim is to find the function, that is, the computational rule, which enables values of h to be calculated from values of t.

We begin with a set of hypothetical experimental results that could apply to the *Tower of Terror* (Table 5.1). We measure the height h in metres above the ground at a sequence of times t, measured in seconds, starting when the car is at the beginning of the vertical part of the tower. We shall ignore all questions of experimental error. These figures, in fact, apply to a body which has been fired into the air. The motion will thus be quicker than that of the car in the *Tower of Terror*, where friction has to be taken into account. The problem is now to find some way of predicting values of h for different values of t. What, for example, is the value of h when $t = 3.4$?

t	0	1	2	3	4	5	6	7	8	9
h	15.0	54.1	83.4	102.9	112.6	112.5	102.6	82.9	53.4	14.1

Table 5.1: Hypothetical data values for the *Tower of Terror*

There are at least two ways to proceed:

- By using graphical techniques. In practice, we would probably not take such an approach for this problem, because the mathematics and physics needed to find solutions are well understood. However, graphical techniques are applicable to other problems where the theory may not be well understood.

- By developing a formula which expresses the height as a function of time. This may be as a result of some theoretical considerations or it may be purely empirical, that is, we use trial and error to find a formula that fits the known data and use it to calculate any unknown values.

We shall explore the graphical method first. The diagram on the left-hand side of Figure 5.1 shows a plot of the points (t, h) in Table 5.1. In the right-hand diagram we have drawn a smooth curve through these points.

We can use the curve in Figure 5.1 to read off values of h for values of t other than those in Table 5.1. This procedure is not entirely satisfactory for a number of reasons:

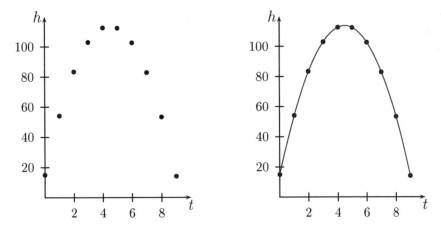

Figure 5.1: Hypothetical data values for the *Tower of Terror*

- We have to make some assumptions about the general shape of the interpolating curve. For instance, why don't we simply join the dots in Figure 5.1 with straight lines?

- There is an inherent inaccuracy in reading figures from a graph, even if the graph is in some sense correct.

- We are making assumptions about the behaviour of the curve in regions where we have no data. See, for example, Figure 3.4 on page 29.

- The results only apply to this particular case. If we wished to calculate height values for some other projectile, we would have to start again from a series of measurements. Obviously, it would be preferable to have a method of calculation that relies on only a few input parameters, such as the initial velocity.

Let us now try to find a formula for the height in terms of the elapsed time. It is easy to find a function represented by a graph when the graph is a straight line, but not in other cases. It would help considerably if we could find some other quantity to plot that would give a straight line. One possibility for such a quantity is the velocity at different times, which we would naturally expect to be an important factor. A faster initial velocity should result in a greater maximum height. In the present problem, the slope of the chord joining two points gives the value of the average velocity between the two times and the slope of the tangent gives the instantaneous velocity. As an experiment we try plotting the instantaneous velocity against time. This may or may not be useful. The velocity can be calculated by estimating the slope of the tangent at a number of points. For example, from Figure 5.2 we can calculate the velocity at $t = 3$ to be about 15 m/s. Of course, if we know the equation of the curve, then we can calculate the velocity by differentiation, but in this case it is the equation that we are trying to find.

Table 5.2 gives the velocity for various times t, where the velocity v is estimated to the nearest 1 m/s.

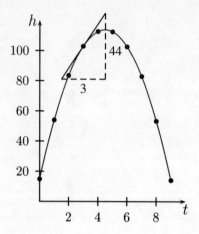

Figure 5.2: The velocity as the slope of the tangent

t (seconds)	1	2	3	4	5	6	7	8
v (m/s)	35	24	15	5	-5	-14	-24	-34

Table 5.2: Velocity at different times

In Figure 5.3, we have plotted these points on a graph. The points lie almost exactly on a straight line. We draw what appears to be the best straight line through the plotted points and then find its equation. The intercept on the v–axis is 44 and the slope is found to be -9.8, so the equation is $v(t) = 44 - 9.8t$. If we wish to be more accurate, we can make use of the *Mathematica* command Fit which will derive the parameters for the best straight line through two or more plot points.

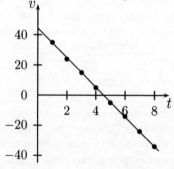

Figure 5.3: The velocity of the projectile

Recall that we are seeking to express the height h in terms of t. Since the velocity is given by the equation

$$\frac{dh}{dt} = v(t),$$

we have
$$h'(t) = 44 - 9.8t. \tag{5.1}$$

This is an equation containing the derivative of the function we need, and we have to use this equation to find the function. Note carefully that the unknown quantity h in equation (5.1) is a *function*, and not a number. Equations such as (5.1), where the unknown is a function and the derivatives of the unknown function appear, are called *differential equations*. The function h is called the *solution* of the differential equation. Differential equations are at the heart of the applications of the calculus and we shall encounter many of them. The problem is now to solve the differential equation we have obtained, that is, we have to find the function h. We will address this problem in the next section.

In obtaining Table 5.2 from Table 5.1, we had to graphically estimate the values of the derivative. There is also a numerical method of approximating derivatives from a given table of values. This is the method of *differences*. For $x_1 < x_2 < x_3$, let (x_1, y_1), (x_2, y_2) and (x_3, y_3) be three tabulated values connecting the dependent variable y with the independent variable x. Then

(a) The method of *forward differences* approximates the derivative $y'(x_2)$ by
$$y'(x_2) \approx \frac{y_3 - y_2}{x_3 - x_2}.$$

(b) The method of *backward differences* approximates the derivative $y'(x_2)$ by
$$y'(x_2) \approx \frac{y_2 - y_1}{x_2 - x_1}.$$

(c) The method of *central differences* approximates the derivative $y'(x_2)$ by
$$y'(x_2) \approx \frac{y_3 - y_1}{x_3 - x_1}.$$

EXAMPLE 5.1

A quantity α is measured at times $t = 0, 1, 2, 3, 4, 5$. Calculate the derivative of α at $t = 3$ by using

(a) the method of forward differences

(b) the method of central differences.

t	0	1	2	3	4	5
$\alpha(t)$	0	0.2	0.8	1.4	3.2	5

(a) The method of forward differences takes the slope of the line joining the point $(3, 1.4)$ to the next point $(4, 3.2)$ as an approximation to the derivative at $t = 3$. Thus
$$\alpha'(3) \approx \frac{3.2 - 1.4}{4 - 3} = 1.8.$$

(b) The method of central differences takes the slope of the line joining the point $(2, 0.8)$ to the point $(4, 3.2)$ as an approximation to the derivative at $t = 3$. Thus
$$\alpha'(3) \approx \frac{3.2 - 0.8}{4 - 2} = 1.2.$$

Central differences take the average of the slope of the chord on either side of the point in question and are usually more accurate.

EXERCISES 5.1

1. You are conducting an experiment in which you measure a quantity α at different times t.

t	0	0.1	0.2	0.3	0.4	0.5	0.6	0.7	0.8	0.9	1.0
$\alpha(t)$	0.1	0.103	0.112	0.127	0.148	0.175	0.208	0.247	0.292	0.343	0.4

   (a) Use the table to calculate $\alpha'(t)$ approximately, for $t = 0.1, 0.2, 0.3, 0.4, 0.5$.

   (b) Plot the graph of $\alpha'(t)$ on the interval $[0.1, 0.5]$, using your approximate results in part (a), above.

   (c) Use your graph to write down a differential equation for $\alpha(t)$.

   (d) Deduce a formula for $\alpha(t)$. Check your formula for the case $t = 0.9$.

2. Tests on the C5 Chevy Corvette gave the results in the table below.

Time, t	0	3	6	9	12
Velocity, v, (m/s)	0	20	33	43	51

   (a) Estimate the acceleration for $t = 3, 6, 9$.

   (b) What can you say about the sign of $v'(t)$ over the period shown? Justify your answer.

   (c) If the car were able to maintain its acceleration at $t = 3$ for the whole period, how fast would it be going after 12 sec?

5.2 SOLVING DIFFERENTIAL EQUATIONS

One general method of solving a differential equation is to make a reasonable guess as to the form of the solution and then substitute this guess into the original equation. This is referred to as *using a trial solution*, but is really just another name for intelligent guessing. Let us try this for the equation

$$h'(t) = 44 - 9.8t. \tag{5.2}$$

We want to find an expression whose derivative is $44 - 9.8t$. This is a polynomial of degree 1 and, since differentiation reduces the degree of a polynomial by 1, it seems reasonable to take a polynomial of degree 2 as a trial solution. Let us take $h(t) = at^2 + bt + c$ as the trial solution to substitute into equation (5.2). We have

$$\frac{d}{dt}(at^2 + bt + c) = 44 - 9.8t,$$

that is,

$$2at + b = 44 - 9.8t. \tag{5.3}$$

Comparing the two sides of this equation enables us to deduce that $a = -4.9$ and $b = 44$. Here we are making use of the result that if two polynomials are equal for all values of t, then the coefficients

of corresponding powers are equal. The justification for this procedure is provided by Theorem 5.1 on page 64. Hence
$$h(t) = -4.9t^2 + 44t + c. \tag{5.4}$$
There is still the constant c to find before we know h. Unfortunately, we cannot get the value of this constant from the differential equation. The reason for this is that the information we get from equation (5.2) is the slope of the tangent to the graph of $f(t)$ at the point t and there are many different curves with the same slope (Figure 5.4). As we shall see in a moment, it is the initial height which determines which curve to select for a given case. We would have obtained the same differential equation if we had started the motion at a height of say $h = 20$ m or $h = 10$ m or $h = 0$ m.

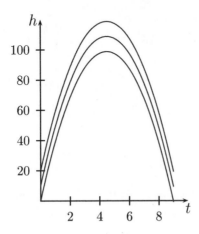

Figure 5.4: Solutions for different initial heights

In the present case $h = 15$ when $t = 0$, that is, $h(0) = 15$. This information, together with equation (5.2) can be written as an *initial value problem*
$$h'(t) = 44 - 9.8t, \quad h(0) = 15,$$
where we give both a differential equation and a value of the dependent variable corresponding to a given value of the independent variable. From equation (5.4), we have
$$15 = h(0)$$
$$= \left.(-4.9t^2 + 44t + c)\right|_{t=0}$$
$$= c,$$
so that $c = 15$ and the solution to the above initial value problem is
$$h(t) = -4.9t^2 + 44t + 15.$$
This equation should reproduce the values in Table 5.1 if our analysis is correct. For instance, if $t = 5$, then
$$h(5) = -4.9 \times 5^2 + 44 \times 5 + 15$$
$$= 112.5,$$

which agrees with the result in Table 5.1. We leave you to check the rest of Table 5.1 as an exercise. Of course, this is not conclusive evidence that we have found the right function. It may be that if we were to measure the height for a value of t other than one in Table 5.1, we might find that equation (5.2) did not agree with the experimental result. This would indicate that there was an error in our theory and we would have to reconsider the situation. However, until this happens (and it never has for falling bodies), we retain our solution (5.2).

Remark

In finding the values of a and b in equation (5.3), we have made use of the following result.

THEOREM 5.1 Equating coefficients
If the two polynomials
$$p(x) = a_0 x^n + a_1 x^{n-1} + \cdots + a_{n-1} x + a_n$$
and
$$q(x) = b_0 x^n + b_1 x^{n-1} + \cdots + b_{n-1} x + b_n$$
are equal for all values of x, then $a_i = b_i$ for each $i = 0, 1, 2, \ldots, n$.

EXAMPLE 5.2

- If $at^3 + 2t^2 + ct + d = 3t^3 + bt^2 + 4t + 5$ for all t, then $a = 3, b = 2, c = 4$ and $d = 5$.
- If $at^3 + 2t^2 + ct + d = bt^2 + 4t + 5$ then $a = 0, b = 2, c = 4$ and $d = 5$.

EXAMPLE 5.3

Solve the initial value problem
$$\frac{dy}{dt} = t^2 + 3t + 4, \quad y(0) = 1.$$

Solution. We take a trial solution $y = at^3 + bt^2 + ct + d$ and substitute it into the given equation. This gives
$$3at^2 + 2bt + c = t^2 + 3t + 4,$$
so that $a = 1/3, b = 3/2$ and c=4. Thus
$$y = \frac{1}{3}t^3 + \frac{3}{2}t^2 + 4t + d,$$
and using the initial condition $y = 1$ when $t = 0$ gives $d = 1$. Thus the solution is
$$y = \frac{1}{3}t^3 + \frac{3}{2}t^2 + 4t + 1.$$

EXERCISES 5.2

In Exercises 1–4, determine the coefficients $a, b, c, \ldots$

1. $ax^3 + bx^2 + cx = x^2 + 3x + d$
2. $ax^3 + bx^2 + c = 0$
3. $at^3 + 3bt^2 + 3t = t^3 + 3t + c$
4. $at^2 + 2bt + c = 3t^2 + 2t + a$

5. If
$$ax^3 + bx^2 + cx + d = a(x - \alpha)(x - \beta)(x - \gamma),$$
show that
$$\alpha + \beta + \gamma = -\frac{b}{a}$$
$$\alpha\gamma + \beta\alpha + \gamma\beta = +\frac{c}{a}$$
$$\alpha\beta\gamma = -\frac{d}{a}.$$

6. Solve the initial value problem.
$$\frac{dy}{dt} = t^2 + 1, \quad y(0) = 3.$$

7. Solve the initial value problem.
$$h'(t) = t^3 + 2t^2 + 3, \quad h(0) = 0.$$

8. Solve the initial value problem.
$$v'(t) = t^2 + 2t + 4, \quad v(0) = 2.$$

9. A stone is dropped from a 10 storey building. Each storey is 3 metres in height. Observers with synchronised watches are stationed at windows and note the time at which the stone passes the midpoint of their window. The windows are halfway between the floor and ceiling. They compare results and arrive at the following table of the time (in seconds) taken for the stone to reach the midpoint of successive windows.

t	0.55	0.96	1.24	1.46	1.66	1.83	1.91	2.14	2.28	2.41

Use the graphical techniques described in this chapter to determine the equation of motion. It is possible that one or more of the observers may have made an error. Check that your equation adequately reproduces the results in the table and decide which, if any, observers have made errors. Justify your assertion.

5.3 GENERAL REMARKS

The procedure we have used in the previous section gives rise to a number of questions:

- We have obtained the equation (5.1) by experimental means, so it is, at best, a good approximation to reality. Is there any way of getting a more accurate equation?

- Equation (5.1) only applies to the table of experimental results of Table 5.1. A new set of experimental results would require us to repeat the above rather lengthy process. Can we get a general equation which can be easily applied to all situations?

In response to the second question we can try to deduce the physical laws which govern the motion of a freely falling body. Subsequently we can use these laws to find the differential equation in any particular case. Initially, we will simplify the problem by assuming that there is no friction or air resistance. Later in the book, we will show how to take these into account. In the case of a body falling under gravity, Galileo around 1633–1638 was the first person to gain a full understanding of the problem. He used experimental results to make general deductions. The physical law that Galileo discovered is the fact that all freely falling bodies have the same constant acceleration, commonly denoted by the symbol g. He is supposed to have demonstrated this fact by dropping two cannonballs of different weights from the top of the Leaning Tower of Pisa and observing that they hit the ground at the same time, although it is unclear whether or not this story is true. Galileo was also the first person to realise that the path of a freely moving projectile is a parabola.

> It has been observed that missiles and projectiles describe a curved path of some sort: however, no one has pointed out the fact that this path is a parabola.
> Galileo 1638

The acceleration is given by $v'(t)$, so that

$$\frac{dv}{dt} = -g, \tag{5.5}$$

where g is a positive constant whose value has to be determined by experiment. In fact, $-g$ is the slope of the graph in Figure 5.3. Galileo's law states that the velocity-time graph for all bodies falling freely under gravity have the same slope $-g$. Equation (5.5) is another example of a differential equation. The function $v(t)$ which satisfies this differential equation is its solution and to solve the differential equation we must find this function. In the present case this is not difficult. We want a function $v(t)$ whose derivative is $-g$ and it is easy to see that

$$v(t) = -gt + c \tag{5.6}$$

is such a function, where c is any constant. The reason that this constant appears is because the original differential equation (5.5) only gives the slope of the line in equation 5.6. There are many straight lines with the same slope, each corresponding to a different value of c. If we put $t = 0$ in equation (5.6), we see that $c = v(0)$, the velocity at time $t = 0$. That is, c is the *initial velocity*, commonly denoted by the symbol u. Note that u is a *number*, whereas v is a *function*.

Since $v = h'(t)$, we have another differential equation

$$h'(t) = -gt + u. \tag{5.7}$$

Equation (5.7) corresponds to the experimental result of equation (5.1) with $g \doteq 9.8 \, \text{ms}^{-2}$ and $u = 44$ m/s. More accurate measurements give $g \doteq 9.81 \, \text{ms}^{-2}$. In order to solve this differential equation, we have to find a function $h(t)$ whose derivative is $-gt + u$. We take $h(t) = at^2 + bt + c$ as a trial solution and quickly find that

$$h(t) = -\frac{1}{2}gt^2 + ut + h_0, \tag{5.8}$$

where $h_0 = h(0)$ is the initial height.

We can now use our solution to compute various properties of the motion. This illustrates the second of the two main uses of calculus. The first use is to find functions which govern the variables in practical problems. The second use is to investigate the properties of these functions. Some of the properties we could be interested in are the following:

- How long does the motion take?

- What is the maximum height reached?

In the case of the *Tower of Terror* problem, the initial height is 15 m and so to answer the first question, we need to find the times when the body is 15 m high. The first time occurs when the body is on the way up and the second time when it is on the way down. From equation (5.8) with $h_0 = 15$, we have
$$15 = 15 + 44t - 4.9t^2,$$
and this is a simple quadratic equation with solutions $t = 0$ s, corresponding to the starting point and $t = 8.98$ s, corresponding to the time that the body is back at 15 m. Thus the total time to go up and down is $t = 8.98$ s. This is less than the actual time taken for *Tower of Terror*, because our hypothetical data in Table 5.1 did not take friction into account.

The second question is also quite easy to answer. As the body rises, its velocity decreases until the maximum height is reached, at which time the instantaneous velocity is zero, that is, $h'(t) = 0$. Thereafter, the velocity becomes increasingly negative. To find the maximum height, we find the time when the velocity is zero and then use equation (5.8) with this value of the time to find the height. Putting $h'(t) = 0$, $u = 44$ m/s and $g = 9.8$ ms^{-2} into equation (5.7) for the velocity gives
$$44 - 9.8t = 0,$$
so that $t = 44/9.8 = 4.49$ s. Substituting this value of t into equation (5.8) gives the maximum height:
$$\text{maximum height} = h(4.49) = 113.8 \text{ m}.$$
We see that the time taken to reach the maximum height is half the time taken to rise and fall. This seems physically reasonable. The free fall time is 4.49 s. This is a little shorter than for the *Tower of Terror*, which is about 6.5 s, due to friction slowing the motion.

5.3.1 The use of *Mathematica*

Mathematica will solve differential equations and plot their solutions. Let us take the differential equation
$$\frac{dh}{dt} = -gt + u,$$
with initial condition $h(0) = 0$. The basic *Mathematica* instruction to solve differential equations is `DSolve`. The command
```
sol=DSolve[{h'[t]==-g*t+u,h[0]==0},h[t],t]
```
solves the differential equation and stores the solution as `sol`. Notice that we need two equations in the instruction—the differential equation itself and the initial condition. We also have to take care to let *Mathematica* know that the function we want to find is h and that the independent variable is t. This is necessary, because *Mathematica* has no way of knowing which of the quantities g, t or u is the independent variable. We also give the solution a name (`sol`), so that we can make further use of it. *Mathematica* gives the following expression for `sol`:
$$\{\{h[t] \to \tfrac{1}{2}(-gt^2 + 2tu)\}\}$$

This is not a convenient form for plotting or further computation, but the expression `sol[[1,1,2]]` will extract $\tfrac{1}{2}(-gt^2 + 2tu)$, the formula for the function. To plot the graph we use the instruction

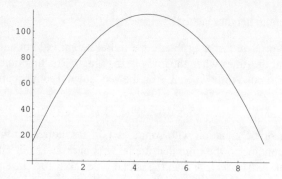

Figure 5.5: *Mathematica* plot of the solution to $h'(t) = -9.8t + 44$

```
Plot[sol[[1,1,2]]/.{g->9.8,u->44},{t,0,10}].
```

This produces the plot shown in Figure 5.5.

EXERCISES 5.3

1. Show that
$$h(t) = -\frac{1}{2}gt^2 + ut + h_0$$
is a solution of the initial value problem
$$h'(t) = -gt + u, \quad h(0) = h_0.$$

2. A ball is thrown vertically upwards. If air resistance is taken into account, would you expect the time during which the ball rises to be longer or shorter than the time during which it falls?

3. A stone is dropped over a cliff. Two seconds later, a second stone is dropped over the cliff. How far apart are the stones 2 seconds after the second one has been dropped?

4. Prove that the difference in the velocities of any two falling bodies is constant.

5. A stone is thrown vertically upwards at 4 m/s. At the same time a second stone is thrown with a velocity which enables it to reach a height of twice that of the first. Find the ratio of times for which the stones are in the air.

6. Use *Mathematica* to solve the initial value problems:

    (a) $\dfrac{dy}{dt} = t^2 + 1, \quad y(0) = 3$

    (b) $h'(t) = t^3 + 2t^2 + 3, \quad h(0) = 0$

    How can you check the correctness of your answers?

5.4 INCREASING AND DECREASING FUNCTIONS

Let's look again at the motion of a body under the influence of gravity. In Figure 5.6, the height function is increasing over the first part of the motion. It reaches a maximum value and then starts to decrease.

This is a particular instance of a general problem. Let f be a given function:

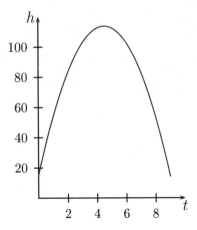

Figure 5.6: The variation of height with time

- Find the portion of its domain where f is increasing.

- Find the portion of its domain where f is decreasing.

- Find a number x in the domain of f for which $f(x)$ achieves a maximum or minimum value.

This type of problem occurs more frequently than one might expect. Here are two instances:

1. The amount of carbon dioxide in the atmosphere is increasing as a result of industrial emissions. This has given rise to the well-known greenhouse effect, which may have serious climatic consequences. The amount of carbon dioxide in the atmosphere is a function of time and is influenced by many factors, including legislation to reduce emissions. Current thinking is that the total amount of atmospheric carbon dioxide should be limited to no more than twice the pre-industrial levels. To ensure that this happens, it is necessary to be able to determine the effect factors such as restrictions on emission rates and reafforestation will have on the maximum value of carbon dioxide in the atmosphere.

2. World population is expected to peak sometime around the middle of the 21st century and thereafter to decline. A good knowledge of the population curve is needed for planning purposes.

Let us clarify the precise meaning of increasing and decreasing functions. Let I be an interval in the domain of a function f.

- The function f is said to be **increasing** on I if for any two points x_1 and x_2 in I with $x_1 < x_2$, we have $f(x_1) < f(x_2)$.

- The function f is said to be **decreasing** on I if for any two points x_1 and x_2 in I with $x_1 < x_2$, we have $f(x_1) > f(x_2)$.

If we look at the graph in Figure 5.6, it is clear that the slope of the tangent is positive when f is increasing. This suggests (but does not prove) that $f'(x)$ will be positive for all $x \in I$ if f is increasing on I. Similarly, we expect $f'(x)$ to be negative for all $x \in I$ if f is decreasing on I. These facts can be proved and we shall assume them to be true. They are summarised in the next theorem.

> **THEOREM 5.2 Increasing and decreasing functions**
> Let f be differentiable on an interval (a,b).
>
> - If $f'(x) > 0$ for all x in (a,b), then f is increasing on (a,b).
> - If $f'(x) < 0$ for all x in (a,b), then f is decreasing on (a,b).

EXAMPLE 5.4
Determine the intervals on which the function $f(x) = x^3 - 6x^2 + 9x$ is

(a) increasing

(b) decreasing.

Solution. We find that
$$f'(x) = 3x^2 - 12x + 9$$
$$= 3(x-1)(x-3).$$

From this we see that $f'(x) > 0$ if $x < 1$ or $x > 3$ and $f'(x) < 0$ if $1 < x < 3$. Consequently, f is decreasing if $1 < x < 3$ and increasing if $x < 1$ or $x > 3$. The graph of f is plotted in Figure 5.7.

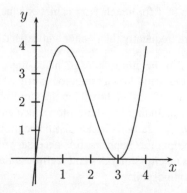

Figure 5.7: The graph of $f(x) = x^3 - 6x^2 + 9x$.

5.5 EXTREME VALUES

Let us now turn to the problem of finding points where a function f takes on extreme values. We shall assume that all our functions are defined on an open interval or on the union of open intervals. We begin with a definition.

> A function f is said to have a **local maximum** at x_0 in its domain if $f(x_0) \geq f(x)$ for all x sufficiently close to x_0. It is said to have a **local minimum** at x_0 in its domain if $f(x_0) \leq f(x)$ for all x sufficiently close to x_0.

A local maximum or local minimum value of a function is often referred to as a *local extreme value* of the function. If $f(x_0)$ is a local extreme value of f, the point $(x, f(x_0))$ is called a *turning*

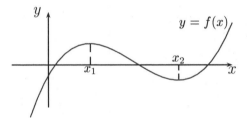

Figure 5.8: Local maxima and minima

point of the graph of f. The function f shown in Figure 5.8 has a local maximum at x_1 and a local minimum at x_2.

The word *local* is used here to indicate that we are only considering function values close to a point x_0 in the domain of the function. In contrast to local extreme values, we can also consider *global* or *absolute* maxima and minima.

> A function f is said to have a **global maximum** at x_0 in its domain if $f(x_0) \geqslant f(x)$ for all x in the domain of f. It is said to have a **global minimum** at x_0 in its domain if $f(x_0) \leqslant f(x)$ for all x in the domain of f.

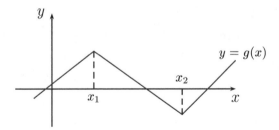

Figure 5.9: A function which is not differentiable at its local extrema

In Figure 5.8, f does not have a global extreme value at either x_1 or x_2. Figure 5.8 suggests that for smooth curves the tangent is horizontal at a point x_0 where f has a local extreme value, in which case $f'(x_0) = 0$. However, if the curve is not smooth, then there may be no derivative at a point where the function has a local maximum or minimum. This is illustrated in Figure 5.9, where the function g has local extreme values at x_1 and x_2, but is not differentiable at these points.

There is a theorem which formalises the above observations. We state it without proof.

> **THEOREM 5.3 Local extreme values**
> *Let f be a function whose domain is an open interval or union of open intervals. If f has a local extreme value at a point x_0 in its domain, then either $f'(x_0) = 0$ or $f'(x_0)$ does not exist.*

A *stationary point* of a function f is a point x_0 for which $f'(x_0) = 0$. Thus local maxima and minima of smooth functions are stationary points. There are, however, other possibilities. A stationary point which is neither a local maximum nor a local minimum is called a *horizontal point of inflexion*. In Figure 5.10, both graphs have a horizontal point of inflexion at x_0.

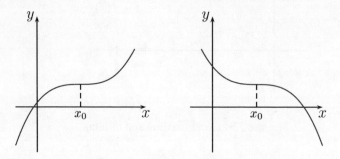

Figure 5.10: Horizontal points of inflexion

To determine whether a stationary point is a local maximum, a local minimum or a horizontal point of inflexion, we examine the sign of the derivative on either side of the stationary point. This is the *first derivative test*.

> **THEOREM 5.4 The first derivative test for maxima and minima**
> *Let f be a function which has a stationary point at x_0 in its domain.*
>
> - *If the sign of $f'(x)$ changes from positive to negative as x passes through x_0, then f has a local maximum value at x_0.*
>
> - *If the sign of $f'(x)$ changes from negative to positive as x passes through x_0, then f has a local minimum value at x_0.*
>
> - *If the sign of $f'(x)$ does not change as x passes through x_0, then f has a horizontal point of inflexion at x_0.*

We shall use the notation x^- to denote a number close to, but less than x, while x^+ will denote a number close to, but larger than x.

EXAMPLE 5.5
Determine the nature of the stationary points of

$$f(x) = x^3 - 6x + 1.$$

Solution. We have $f'(x) = 3x^2 - 6$, so that $f'(x) = 0$ when $x = \pm\sqrt{2}$. Since $f'(\sqrt{2}^-) < 0$ and $f'(\sqrt{2}^+) > 0$, we see that the sign of $f'(x)$ changes from negative to positive as x passes through $\sqrt{2}$. Consequently, f has a local minimum at $\sqrt{2}$. Similarly, there is a local maximum at $-\sqrt{2}$.

5.5.1 Concavity

Consider Figure 5.11, which shows the graph of a function f. On the interval (a, b), the graph, roughly speaking, curves up, while on the interval (b, c), it curves down. The technical terms for this behaviour are *concave up* and *concave down*. In the definition below, we can make these terms precise by noticing that on the interval (a, b), the function f' is increasing, while on (b, c), the function f' is decreasing. (Why?)

EXTREME VALUES

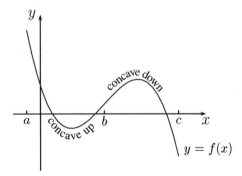

Figure 5.11: Concavity on an interval

> Let f be a function which is differentiable on an open interval I. The graph of f is said to be **concave up** on I if f' is increasing on I. The graph of f is said to be **concave down** on I if f' is decreasing on I.

Suppose that f' is differentiable on an open interval I. It follows that if the graph of f is concave up on I, then $f''(x) > 0$ for all $x \in I$. If the graph of f is concave down on I, then $f''(x) < 0$ for all $x \in I$. A point x_0 where the concavity of the graph of f changes sign is called a *point of inflexion*, in which case $f''(x_0) = 0$. Note carefully that the converse of this result is not true. *The result $f''(x_0) = 0$ does not imply that there is a point of inflexion at x_0.* An example of a point of inflexion occurs in Figure 5.11, where the graph of f has a point of inflexion at $(b, f(b))$.

EXERCISES 5.5

In Exercises 1–4, find the intervals on which f increases and the intervals on which f decreases.

1. $f(x) = x^3 - 3x + 2$
2. $f(x) = x^3(1 + x)$
3. $f(x) = x^2 + x - 30$
4. $f(x) = |2x + 7|$

In Exercises 5–8, find the stationary points of f and the local extreme values.

5. $f(x) = 4x^2 - 8$
6. $f(x) = x^4 - 18x^2$
7. $f(x) = |x^2 - 16|$
8. $f(x) = \dfrac{1 + x}{1 - x}$

9. Find the extreme values of the function $f(x) = x^2 - 4x + 1$, $0 \leqslant x \leqslant 3$.

10. A rectangle has a perimeter of $24\,\text{cm}$. Find the dimensions of the sides that will give the largest area.

11. A beam with a rectangular cross section is to be cut from a circular log of radius $15\,\text{cm}$. The strength of such a beam is proportional to bh^2, where b is its breadth and h its width. Find the dimensions of the strongest beam that can be obtained.

12. An open gutter is to be made from sheet metal of width $22\,\text{cm}$ by folding up identical strips on each side, so that they are at right angles to the base. Determine the dimensions of the gutter that will have the greatest carrying capacity.

CHAPTER 6

SERIES AND THE EXPONENTIAL FUNCTION

In Chapter 5, we solved the problem of bodies moving freely under the influence of gravity by considering some experimental results, using these results to obtain a differential equation and then solving this equation to find the required function that expresses height in terms of time. This same procedure will be used to deal with the second problem, but we shall require a new way of defining functions in order to find the solution.

6.1 THE AIR PRESSURE PROBLEM

The measurement of air pressure is not a simple matter, because air pressure at any given height varies with time. Indeed, this variation is one of the main factors determining weather conditions. Technically, air pressure is a function of two variables, the height h and the time t. We shall not be considering functions with two independent variables in this book and so we shall make some simplifying assumptions.

The SI unit of pressure is the pascal (Pa), which is the pressure exerted by a cylinder of cross section 1 m^2 and height sufficient to give it a mass of 1 kg. However, for historical reasons air pressure is measured in hectopascals (hPa), where one hectopascal is 100 pascals. At sea level, the air pressure varies between 980 and 1030 hectopascals, depending on the weather conditions, but the average value is 1013 hectopascals. Meteorologists usually express the pressure in millibars (mb), where 1 mb is equivalent to 1 hPa.

Instead of bringing the time variation into our considerations, we shall look at the *average* pressure at each height. By using this average we obtain the values given in Table 6.1. In this table, we denote the pressure by p and the height by h. Height is the independent variable and is given in steps of 1000 m up to 10 000 m, which is about the maximum height of a commercial airliner.

h	0	1000	2000	3000	4000	5000	6000	7000	8000	9000	10 000
p	1013	899	795	701	616	540	472	411	356	307	264

Table 6.1: Average air pressure

Plotting p against h gives the curve shown in Figure 6.1.

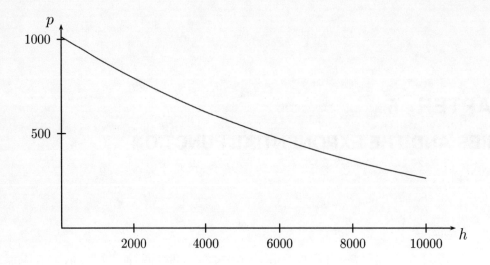

Figure 6.1: Variation of air pressure with height

Here the problem is to find the rule for calculating values of p for given values of h, so that we need a function f which enables us to express the pressure p in the form $p = f(h)$. In what follows, it will be more convenient to abuse the notation and take the approach we did in Chapter 5 (page 58). We will write $p(h)$ instead of $f(h)$. The symbol p is thus doing double duty. It may denote either a number or the rule for calculating this number. The function $p(h)$ is expected to be smooth, so we might try the same approach as we did in Chapter 5.

We plot dp/dh against h and see if we get a straight line. We cannot compute p' from the given data, but we can approximate it by measuring slopes of the graph in Figure 6.1. We have done this by taking *central differences*, as described on page 61. We group the plot points in threes and take the slope of the chord between the first and the third point as an approximation to the slope of the tangent at the middle point. If we do this, we obtain the following data.

h	0	1000	2000	3000	4000	5000	6000	7000	8000	9000	10 000
$\dfrac{dp}{dh}$	-0.12	-0.11	-0.099	-0.089	-0.080	-0.071	-0.064	-0.058	-0.052	-0.046	-0.042

Table 6.2: The derivative of the pressure

If we plot the graph from this table, we get another curve rather than a straight line (Figure 6.2), so the approach of Chapter 5 fails in this case.

What do we try next? Let's think about it. The pressure depends on the mass of the column of air above a given point and as we go higher the air gets thinner because air is compressible. This means that as we go higher, a given volume of air weighs less and so its contribution to the overall pressure is less important. This suggests that when the pressure is high, the rate of change of pressure will also be high and, as the pressure decreases, the rate of change should also decrease. It also suggests that the rate of change of pressure depends basically on the pressure rather than the height.

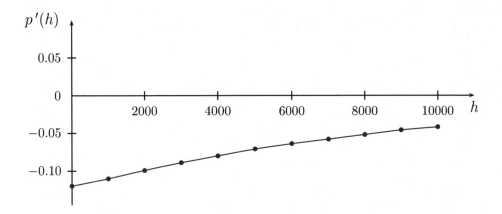

Figure 6.2: Variation of the derivative of the air pressure with height

Given this argument, we try plotting p' against p and see what happens. We don't know in advance if this will achieve anything, but it illustrates the way in which we must use all our physical intuition about a given situation in trying to produce a workable mathematical model of it. In this case we do indeed get a straight line (Figure 6.3) to within a small amount of error.

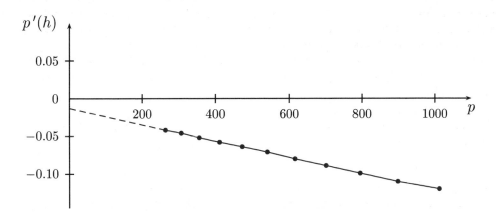

Figure 6.3: Variation of the derivative of the pressure with pressure

The equation of this line can be obtained and this equation, together with the initial value $p(0) = 1013$, gives us the initial value problem

$$p'(h) = -0.000108 p(h) - 0.013, \quad p(0) = 1013. \tag{6.1}$$

It is this initial value problem that has to be solved for the pressure as a function of height. It is more complicated than those we encountered in Chapter 5 for two reasons. First, the constants are less friendly than the previous cases. More importantly, the function p as well as its derivative appears in the equation. This seemingly minor matter actually complicates the solution considerably. Rather than attack equation (6.1) directly, we consider a simpler form. This is often a useful procedure in difficult problems. Examine an easier version of the problem first in order to gain the necessary insight before attempting the more difficult problem. Equation (6.1) has the form

$$p'(h) = a + bp(h), \quad p(0) = c,$$

where a, b and c are constants. We shall consider the special case $a = 0$, $b = 1$ and $c = 1$. That is, we consider the initial value problem

$$p'(h) = p(h), \quad p(0) = 1.$$

We now have a problem which is purely mathematical, namely the problem of finding p from this equation. In words, this problem is to find a function which is equal to its own derivative. This is a typical instance of the way mathematical and scientific ideas develop in tandem. A physical problem gives rise to a mathematical problem and subsequent investigation of the mathematical problem may shed new light on the physical problem. We take a general approach and put $y = f(x)$ and, if y is equal to its own derivative, then

$$\frac{dy}{dx} = y, \; y(0) = 1. \tag{6.2}$$

To solve this problem we need to find y as a function of x. There is more than one way of doing this and we shall consider two methods in this book, one in this chapter and another in Chapter 10.

We begin by trying the approach we developed in Chapter 5, where we solved differential equations by using polynomial functions as trial solutions. Although this will not work here for reasons which will soon become clear, it is still useful to see what happens if we attempt to use polynomials as trial solutions. We find out why they don't work and what we have to do to actually get the solution we are seeking. The methods developed will prove to be useful in later problems.

Take a trial solution $y = ax + b$, where a and b are constants. If we substitute this into the first equation of (6.2), the left hand side becomes

$$\frac{dy}{dx} = a$$

while the right hand side is

$$y = ax + b.$$

We then have $a = ax + b$ and so equating coefficients gives $a = 0$ and $b = a$. Thus $y(x) = 0$ for all values of x. This is a solution of the differential equation, but does not satisfy the initial condition $y(0) = 1$. If we try a higher degree polynomial, a similar problem arises. Suppose, for example, we try $y = ax^3 + bx^2 + cx + d$. Then the left hand side of the first equation in (6.2) becomes

$$\frac{dy}{dx} = 3ax^2 + 2bx + c$$

while the right hand side is

$$y = ax^3 + bx^2 + cx + d.$$

We then have

$$3ax^2 + 2bx + c = ax^3 + bx^2 + cx + d. \tag{6.3}$$

Once again, there is a discrepancy between the degrees of the terms on the two sides of the equation— the left hand side has degree 2, while the right hand side has degree 3. If we equate coefficients

(Theorem 5.1), we get

$$c = d \quad \text{(constant terms)}$$
$$b = \frac{c}{2} = \frac{d}{2} \quad \text{(coefficient of } x\text{)}$$
$$a = \frac{b}{3} = \frac{c}{3.2} = \frac{d}{3.2} \quad \text{(coefficient of } x^2\text{)}$$
$$a = 0 \quad \text{(coefficient of } x^3\text{)}$$

There is a promising pattern emerging which is unfortunately spoiled by the fact that there are not enough terms on the left hand side of equation (6.3) leading to the result $a = 0$. If $a = 0$, it quickly follows that $b = c = d = 0$ and we cannot satisfy the initial conditions. This problem of running out of terms on one side of the equation will occur whenever we take a trial solution of the form $y(x) = a_0 + a_1 x + a_2 x^2 + \cdots + a_{n-1} x^{n-1} + a_n x^n$. We will find that $a_k = 0$, $k = 0, 1, 2, \ldots, n$.

However, suppose we use a trial solution which never runs out of terms, that is, suppose we try a solution of the form

$$y(x) = a_0 + a_1 x + a_2 x^2 + \cdots + a_n x^n + \ldots, \tag{6.4}$$

where the terms go on forever. The right hand side of equation (6.4) is an infinite sum, rather than a finite sum. Such an infinite sum is called an *infinite series* and we shall soon have to be quite precise about the meaning and definition of an infinite series, but for the moment let us press on with trying to find a solution to the initial value problem (6.2). If we substitute equation (6.4) into equation (6.2) we find

$$a_1 + 2a_2 x + 3a_3 x^2 + \cdots + na_n x^{n-1} + (n+1)a_{n+1} x^n + \ldots$$
$$= a_0 + a_1 x + a_2 x^2 + \cdots + a_n x^n + \ldots \tag{6.5}$$

We have assumed that we can differentiate equation (6.4) term by term to get equation (6.5). This is an assumption which we shall have to examine later. Equating coefficients in this equation gives us

$$a_1 = a_0$$
$$a_2 = \frac{a_1}{2} = \frac{a_0}{2}$$
$$a_3 = \frac{a_2}{3} = \frac{a_1}{3.2} = \frac{a_0}{3.2}$$
$$\vdots$$
$$a_n = \frac{a_{n-1}}{n} = \cdots = \frac{a_0}{n!}$$
$$\vdots$$

so that

$$y(x) = a_0 \left(1 + x + \frac{x^2}{2} + \frac{x^3}{3!} + \cdots + \frac{x^n}{n!} + \ldots \right) \tag{6.6}$$

Finally, we evaluate the constant a_0 by using the fact that $y = 1$ when $x = 0$ (equation (6.2)). Putting $x = 0$ in each side of equation (6.6) gives $a_0 = 1$, leading to

$$y(x) = 1 + x + \frac{x^2}{2} + \frac{x^3}{3!} + \cdots + \frac{x^n}{n!} + \ldots \tag{6.7}$$

At this stage it is not clear that we can get any useful information out of equation (6.7). How can we operate sensibly with an infinite number of terms? We certainly cannot add up an infinite number of terms, because the additions go on forever. In part, the answer lies in the fact that in any practical calculation we only want accuracy to a certain number of decimal places. Notice that in equation (6.7), the denominators of the terms become larger as we move to the right, so that for a fixed value of x, we would expect that eventually successive terms will become smaller and smaller. What we hope is that they become so small that we can ignore all terms after a certain number and still attain the accuracy we want. In the case of equation (6.7) this is indeed what happens. We can illustrate (but not *prove*) this in the following table. We have used *Mathematica* to calculate the *finite* sum

$$y(x) = 1 + x + \frac{x^2}{2} + \frac{x^3}{3!} + \cdots + \frac{x^n}{n!}$$

for three values of x and various increasing values of n.

	$n=0$	$n=1$	$n=5$	$n=10$	$n=15$	$n=50$	$n=100$
$x=1$	1	2	2.70823	2.71828	2.71828	2.71828	2.71828
$x=5$	1	6	65.375	143.689	148.38	148.413	148.413
$x=10$	1	11	644.333	10086.6	20188.2	22026.5	22026.5

Table 6.3: Partial sums $1 + x + \frac{x^2}{2!} + \cdots + \frac{x^n}{n!}$ for various values of n and x

Notice how, for a given x, the terms all appear to approach a fixed value as n becomes progressively larger. The technical term for this phenomenon is known as *convergence*. A table such as this is not conclusive, but it does suggest that if we fix a value of x, then by taking enough terms of the series (6.7) we can calculate $y(x)$ to any accuracy we want. In the next section, we will consider infinite series in more detail, with a view to justifying these assertions.

The calculations we have carried out here are all exploratory and suggestive of methods which might be used to find a solution of the initial value problem $y'(x) = y(x)$, $y(0) = 1$. Our task now is to investigate how to make them precise. We have to answer questions such as:

- How do we add up an infinite sum?

- Can we differentiate an infinite sum of functions in the same way as a finite sum?

- Is there a function, defined as an infinite sum, which is a solution of the differential equation?

EXERCISES 6.1

1. The following table gives the density ρ of the atmosphere at a height h, for different values of h. The density is given as a fraction of the mean sea level density, while the height is given in kilometres. Based on this data, determine a differential equation relating the height and the density.

ρ	0.5	0.25	0.125	0.0625	0.0312	0.0156	0.0078	0.0039	0.0019
h	5.28	10.56	15.84	21.12	26.40	31.68	36.96	42.24	47.52

6.2 INFINITE SERIES

We begin with a provisional definition:

> **Provisional definition 6.1** *A* **series** *is an infinite sum of numbers. The numbers which appear in the sum are called the* **terms** *of the series.*

There are two types of notation for series. The first is to write down the first few terms of the series and use dots to indicate that the sum does not terminate. This is the way we have written series in the discussion above. The second is to use *summation notation*. The symbol $\sum_{k=1}^{n} a_k$ is used to denote the finite sum $a_1 + a_2 + \cdots + a_n$, while $\sum_{k=1}^{\infty} a_k$ is used to denote the infinite sum $a_1 + a_2 + a_3 + \cdots$. We may replace a_k by a formula involving k in both the above cases.

EXAMPLE 6.1
Some particular series are:

- $1 + \dfrac{1}{2} + \dfrac{1}{4} + \dfrac{1}{8} + \cdots$ or $\sum_{k=1}^{\infty} \dfrac{1}{2^{k-1}}$

- $1 + 2 + 3 + 4 + \cdots$ or $\sum_{k=1}^{\infty} k$

- $1 + \dfrac{1}{2} + \dfrac{1}{3} + \dfrac{1}{4} + \cdots$ or $\sum_{k=1}^{\infty} \dfrac{1}{k}$

- $1 - \dfrac{1}{2} + \dfrac{1}{3} - \dfrac{1}{4} + \cdots$ or $\sum_{k=1}^{\infty} \dfrac{(-1)^{k+1}}{k}$ □

It is somewhat misleading to say that a series is an infinite sum, because there is no way to add up an infinite number of terms in the same way that we can do for a finite sum. To say what is meant by the sum of a series requires the theory of sequences which we discussed in Chapter 2. We recall the following facts on sequences:

- A *sequence* is an infinite list of numbers. The numbers in the list are called the *terms* of the sequence.

- A sequence $\{a_n\}$ is said to be *convergent* if $\lim_{n \to \infty} a_n$ exists. Otherwise the sequence is said to be *divergent*.

It is essential to be clear about the difference between sequences (lists) and series (sums). It should be noted that in everyday usage, sequences are almost always referred to as series. For example, a list of stock market prices is referred to as a series of prices.

There are two sequences associated with every series. The first is the sequence of terms. For example, in the series
$$\sum_{k=1}^{\infty} 1/(2^{k-1}) = 1 + \frac{1}{2} + \frac{1}{4} + \frac{1}{8} + \cdots,$$

the sequence of terms is

$$1, \frac{1}{2}, \frac{1}{4}, \frac{1}{8}, \ldots$$

The sequence of terms is important in the theory of series, but much more important is the second of the two sequences associated with a series. This is the sequence of *partial sums*.

> **DEFINITION 6.1 Partial sums:**
> *The nth **partial sum** of the series $\sum_{k=1}^{\infty} a_k$ is the sum of the first n terms of the series. It is denoted by S_n and so $S_n = a_1 + a_2 + \cdots + a_n$.*

The partial sums of a series form a sequence $S_1, S_2, S_3, S_4, \ldots$, and it is this sequence which enables us to say what we mean by the sum of a series.

> **DEFINITION 6.2 Sum of a series:**
> *Let $\sum_{k=1}^{\infty} a_k$ be a series and let $\{S_n\}$ be its sequence of partial sums. The series is said to be **convergent** if the sequence $\{S_n\}$ is convergent. If the series is convergent, then $\lim_{n \to \infty} S_n$ is called the **sum** of the series and is denoted by $\sum_{k=1}^{\infty} a_k$. A series which is not convergent is **divergent**.*

It is not usually possible to find a formula for the sum of a convergent series, except in some special cases.

EXAMPLE 6.2
Consider the series

$$\sum_{k=1}^{\infty} x^{k-1} = 1 + x + x^2 + x^3 + \cdots \tag{6.8}$$

The nth partial sum of this series is

$$S_n = \sum_{k=1}^{n} x^{k-1} = 1 + x + x^2 + x^3 + \cdots + x^{n-1}.$$

This is a finite geometric series with sum

$$S_n = \frac{1 - x^n}{1 - x}.$$

To determine the convergence properties of the infinite series, we look at the sequence of partial sums

$$\{S_n\} = \left\{ \frac{1 - x^n}{1 - x} \right\}.$$

There are two cases to consider:

- If $|x| < 1$, that is, $-1 < x < 1$, then $x^n \to 0$ as $n \to \infty$, so that $S_n \to \frac{1}{1-x}$ as $n \to \infty$. Thus if $|x| < 1$ we can write

$$1 + x + x^2 + x^3 + \cdots = \frac{1}{1-x}$$

  for the sum of the series (6.8).

- If $|x| \geqslant 1$, then $\{x^n\}$ diverges and so the sequence of partial sums also diverges. This means that the series (6.8) diverges if $x \geqslant 1$.

Thus if f is the function defined by $f(x) = 1 + x + x^2 + x^3 + \cdots$, then the domain of f is the open interval $(-1, 1)$ and for any $x \in (-1, 1)$,

$$f(x) = \frac{1}{1-x}.$$

□

In the above example the function we have produced is not new. It is just the simple rational function $1/(1-x)$, although when the series form is used for the definition, the domain is reduced from $\mathbb{R} \setminus \{1\}$ to $(-1, 1)$. We shall come to the reason for this later. This example also shows that a series can be an alternative form of a function we already know. However, some series cannot be expressed in terms of these known functions and it is in such cases that our main interest lies. The series

$$y(x) = 1 + x + \frac{x^2}{2} + \frac{x^3}{3!} + \cdots + \frac{x^n}{n!} + \cdots$$

which occurred in our discussion of atmospheric pressure is just such a series. In that discussion we implicitly assumed that the series converged, but this is something we have to verify. If the series does not converge, then our conclusions in Section 6.1 will have to be re-examined.

EXERCISES 6.2

Evaluate the following sums.

1. $\sum_{k=0}^{k=3} (2k+1)$

2. $\sum_{k=0}^{k=4} (2k-1)$

3. $\sum_{k=1}^{k=3} 3^k$

4. $\sum_{n=0}^{n=3} \frac{1}{n!}$

5. $\sum_{i=1}^{i=3} (-1)^i 2^i$

6. $\sum_{j=3}^{j=5} \frac{1}{2j+1}$

7. $\sum_{n=1}^{\infty} \frac{1}{3^n}$

8. $\sum_{n=0}^{\infty} \left(\frac{2^{n+1}}{3^n} \right)$

9. $\sum_{k=1}^{\infty} \frac{3^{k-1} - 1}{6^{k-1}}$

10. $\sum_{n=1}^{\infty} \left(\frac{5}{2^n} - \frac{2}{3^n} \right)$

11. $\sum_{n=4}^{\infty} \frac{1}{4^n}$

12. $\sum_{n=1}^{\infty} (-1)^n \frac{1}{4^n}$

Express the following sums in Σ notation.

13. $1 + 2 + 3 + 4 + 5 + \ldots$

14. $1 + 3 + 5 + 7 + 9 + \ldots$

15. $\frac{1}{3} + \frac{1}{5} + \frac{1}{7} + \cdots + \frac{1}{15}$

16. $\underbrace{1 + 1 + 1 + \cdots + 1}_{n \text{ terms}}$

17. $\frac{1}{2} + \frac{1}{6} + \frac{1}{12} + \frac{1}{20} + \ldots$

18. $3x - 4x^2 + 5x^3 - 6x^4 + 7x^5$

Suppose that $\sum_{k=1}^{n} a_k = 3$ and $\sum_{k=1}^{n} b_k = -4$. Find the values of the following sums.

19. $\sum_{k=1}^{n} 3a_k$

20. $\sum_{k=1}^{n} (a_k + b_k)$

21. $\sum_{k=1}^{n} (2b_k - 3a_k)$

22. $\sum_{k=1}^{n} \frac{a_k}{5}$

23. Show that
$$1 + 2 + 3 + \cdots + n = \frac{n(n+1)}{2}.$$

24. Sum the series
$$\sum_{n=1}^{\infty} \left(\frac{1}{\sqrt{n+1}} - \frac{1}{\sqrt{n+2}} \right).$$

25. Verify that
$$\frac{1}{n(n+1)} = \frac{1}{n} - \frac{1}{n+1}.$$
Hence show that
$$\sum_{n=1}^{\infty} \frac{1}{n(n+1)} = 1.$$

26. Verify that
$$\frac{1}{(n+1)(n+2)} = \frac{1}{n+1} - \frac{1}{n+2}.$$
Hence evaluate
$$\sum_{n=1}^{\infty} \frac{1}{(n+1)(n+2)}.$$

27. Verify that
$$\frac{6}{(2n-1)(2n+1)} = \frac{3}{2n-1} - \frac{3}{2n+1}.$$
Hence evaluate
$$\sum_{n=1}^{\infty} \frac{6}{4n^2 - 1}.$$

6.3 CONVERGENCE OF SERIES

There are a variety of methods for determining whether a given series converges or not. In Example 6.2 we computed the sum of a series by computing a sequence of partial sums and then finding the limit of this sequence of partial sums. Such direct methods are usually not possible except in a few simple cases, so that indirect methods are necessary. One of the most useful and important tests for convergence is the *ratio test* which we state without proof.

> **THEOREM 6.1 The ratio test**
> *Let*
> $$\sum_{k=1}^{\infty} u_k = u_1 + u_2 + u_3 + \cdots \qquad (6.9)$$
> *be a given series of non-zero terms. Suppose that the sequence*
> $$\left|\frac{u_2}{u_1}\right|, \left|\frac{u_3}{u_2}\right|, \left|\frac{u_4}{u_3}\right|, \cdots = \left\{\left|\frac{u_{k+1}}{u_k}\right|\right\}$$
> *converges to a limit ρ.*
>
> 1. *If $\rho < 1$, then the series (6.9) converges.*
> 2. *If $\rho > 1$, then the series (6.9) diverges.*
> 3. *If $\rho = 1$, then no conclusion can be drawn. The series (6.9) may converge or it may diverge.*

Let us apply this test to the function defined by

$$f(x) = 1 + x + \frac{x^2}{2!} + \frac{x^3}{3!} + \cdots. \qquad (6.10)$$

For a particular value of x, the kth term is $x^{k-1}/(k-1)!$ and using the ratio test we have

$$\lim_{k \to \infty} \left|\frac{x^k/k!}{x^{k-1}/(k-1)!}\right| = \lim_{k \to \infty} \left|\frac{x}{k}\right|$$
$$= |x| \lim_{k \to \infty} \frac{1}{k}$$
$$= 0.$$

Since this holds for every value of x, it follows that the series converges for every x and so the domain of f is $\mathbb{R}$. For any $x \in \mathbb{R}$, if we take enough terms of the series, we can compute $f(x)$ to any required accuracy. This raises an immediate question. How do we determine when enough terms have been taken? Let us put this another way. We want to determine n so that

$$1 + x + \frac{x^2}{2} + \frac{x^3}{3!} + \cdots + \frac{x^n}{n!}$$

gives the sum of the series

$$f(x) = 1 + x + \frac{x^2}{2} + \frac{x^3}{3!} + \cdots + \frac{x^n}{n!} + \cdots$$

to within the required accuracy. Let us write

$$f(x) = 1 + x + \frac{x^2}{2} + \frac{x^3}{3!} + \cdots + \frac{x^n}{n!} + \cdots$$
$$= S_n + \frac{x^n}{n!} + \frac{x^{n+1}}{(n+1)!} + \cdots$$
$$= S_n + E_n.$$

This shows that the error made in approximating $f(x)$ by S_n is

$$E_n = \frac{x^n}{n!} + \frac{x^{n+1}}{(n+1)!} + \cdots$$

We can't calculate this exactly, but we can estimate it and as long as we take care to overestimate it, we can be sure that we have computed $f(x)$ to within the required degree of accuracy. For positive values of x we have

$$\begin{aligned} E_n &= \frac{x^n}{n!}\left(1 + \frac{x}{n+1} + \frac{x^2}{(n+1)(n+2)} + \cdots\right) \\ &\leqslant \frac{x^n}{n!}\left(1 + \frac{x}{n} + \frac{x^2}{n^2} + \frac{x^3}{n^3} + \cdots\right) \\ &= \frac{x^n}{n!}\frac{1}{1 - x/n}, \text{ if } |x| < n, \\ &= \frac{x^n}{(n-1)!(n-x)} \end{aligned} \qquad (6.11)$$

EXAMPLE 6.3

Suppose we want to compute $f(\frac{1}{2})$ for the function

$$f(x) = 1 + x + \frac{x^2}{2} + \frac{x^3}{3!} + \cdots + \frac{x^n}{n!} + \cdots$$

with an error of less than $\varepsilon = 0 \cdot 00001$. Using inequality (6.11) with $x = \frac{1}{2}$, we find

$$E_n \leqslant \frac{1/2^n}{(n-1)!(n-1/2)} < \begin{cases} 0 \cdot 00003 & \text{if } n = 6 \\ 0 \cdot 000002 & \text{if } n = 7 \end{cases}$$

Thus we use $n = 7$ and compute

$$f\left(\frac{1}{2}\right) \approx 1 + \frac{1}{2} + \cdots + \frac{1}{2^7 7!} \doteq 1.64872$$

with an error of less than ε. □

If $x < 0$ in equation (6.10), it is a bit easier to compute the error. The signs of the terms in the series alternate and when this occurs for a convergent series in which the terms are decreasing in absolute value, it can be shown that the error made by taking a given number of terms is always smaller (in absolute value) than the first omitted term. So if $x < 0$, compute $x^n/n!$ until we get a result smaller than the allowable error ε. Then

$$f(x) \approx 1 + x + \frac{x^2}{2} + \frac{x^3}{3!} + \cdots + \frac{x^{n-1}}{(n-1)!}$$

with an error of less than ε. At this stage you should re-examine Example 2.16 of Chapter 2 in the light of the above discussion.

The series

$$y(h) = 1 + h + \frac{h^2}{2} + \frac{h^3}{3!} + \cdots + \frac{h^n}{n!} + \cdots = \sum_{n=0}^{\infty} \frac{h^n}{n!} \qquad (6.12)$$

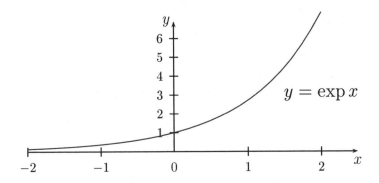

Figure 6.4: The exponential function

has been shown to converge for each value of h. We now reach a very important definition.

> **DEFINITION 6.3 The exponential function:**
> *The **exponential function** $\exp(x)$ is defined by*
> $$\exp(x) = 1 + x + \frac{x^2}{2} + \frac{x^3}{3!} + \cdots + \frac{x^n}{n!} + \cdots = \sum_{n=0}^{\infty} \frac{x^n}{n!}.$$

For reasons which will become clear later on, we often write e^x for $\exp(x)$, where e is a constant whose value is $e = 2.71828\ldots$ The domain of the exponential function is $\mathbb{R}$ and by calculating $\exp(x)$ for various values of x, we can plot its graph, as in Figure 6.4. The value of $\exp(0)$ is easily seen to be 1.

We were led to the exponential function because of our interest in the initial value problem $y'(x) = y(x)$, $y(0) = 1$. To show that $\exp(x)$ satisfies this initial value problem is, in principle, a simple matter. We have to show that

$$\frac{d}{dx}\exp(x) = \exp(x), \quad \exp(0) = 1.$$

The first of these equations is the one that will cause concern. In order to evaluate

$$\frac{d}{dx}\exp(x),$$

we have to know how to differentiate the series which defines the exponential function. More generally, we would like to know how to differentiate any series, assuming that such a series is differentiable. In going from equation (6.4) to equation (6.5), we assumed that we could differentiate the series term by term. This is indeed true in this case, but is not true for series in general. We need to be precise about what we can and cannot do when differentiating series. This will be discussed in the next section. We conclude this section with some additional examples on the use of the ratio test.

EXAMPLE 6.4
Determine whether or not the series
$$\sum_{n=0}^{\infty} \frac{1}{n!}$$
converges.

We have
$$u_n = \frac{1}{n!}, \quad u_{n+1} = \frac{1}{(n+1)!}.$$

Hence
$$\lim_{n \to \infty} \left| \frac{u_{n+1}}{u_n} \right| = \lim_{n \to \infty} \left| \frac{1}{(n+1)!} \times \frac{n!}{1} \right|$$
$$= \lim_{n \to \infty} \left| \frac{1}{(n+1)n!} \times \frac{n!}{1} \right|$$
$$= \lim_{n \to \infty} \left| \frac{1}{(n+1)} \right|$$
$$= 0.$$

Since this is less than 1, the series will converge.

EXAMPLE 6.5
In the series
$$\sum_{n=0}^{\infty} \frac{2^n}{n^2}$$
we have
$$u_n = \frac{2^n}{n^2}, \quad u_{n+1} = \frac{2^{n+1}}{(n+1)^2}.$$

Hence
$$\lim_{n \to \infty} \left| \frac{u_{n+1}}{u_n} \right| = \lim_{n \to \infty} \left| \frac{2^{n+1}}{(n+1)^2} \times \frac{n^2}{2^n} \right|$$
$$= \lim_{n \to \infty} \left| \frac{2n^2}{(n+1)^2} \right|$$
$$= 2.$$

Since this is greater than 1, the series will diverge.

EXAMPLE 6.6
In the series
$$\sum_{n=0}^{\infty} \frac{1}{2n-1}$$
we have
$$u_n = \frac{1}{2n-1}, \quad u_{n+1} = \frac{1}{2n+1}.$$

Hence

$$\lim_{n\to\infty}\left|\frac{u_{n+1}}{u_n}\right| = \lim_{n\to\infty}\left|\frac{1}{2n+1} \times \frac{2n-1}{1}\right|$$

$$= \lim_{n\to\infty}\left|\frac{2n-1}{2n+1}\right|$$

$$= \lim_{n\to\infty}\left|\frac{2-1/n}{2+1/n}\right|$$

$$= \frac{2}{2}$$

$$= 1$$

Since this is equal to 1, we can draw no conclusion from the ratio test.

EXERCISES 6.3

For each of the following series, calculate the value of the partial sum S_3 correct to four decimal places and use the ratio test to decide whether or not the series converges.

1. $\sum_{n=1}^{\infty} \frac{2n+5}{3^n}$

2. $\sum_{n=1}^{\infty} (-1)^{n+1} \frac{(2n+1)!}{n}$

3. $\sum_{n=1}^{\infty} \frac{n!}{(2n)!}$

4. $\sum_{n=1}^{\infty} \frac{n^{10}}{2^n}$

5. $\sum_{n=1}^{\infty} (-1)^{n+1} \frac{5^n}{4^n+5}$

6. $\sum_{n=1}^{\infty} \frac{10^n+1}{(1.1)^n}$

7. $\sum_{n=1}^{\infty} \frac{n!}{10^n}$

8. $\sum_{n=1}^{\infty} \frac{(n!)^2}{(2n)!}$

9. $\sum_{n=1}^{\infty} \frac{1\cdot 3\cdot 5\cdot \ldots \cdot (2n-1)}{n!}$

10. $\sum_{n=1}^{\infty} \left(\frac{2}{n}\right)^n$

11. Show that
$$e^{-x} = 1 - x + \frac{x^2}{2!} - \frac{x^3}{3!} + \cdots$$

12. In 1734, the Swiss mathematician Leonhard Euler proved the following result.
$$1 + \frac{1}{2^2} + \frac{1}{3^2} + \frac{1}{4^2} + \frac{1}{5^2} + \cdots = \frac{\pi^2}{6}.$$

This is a bizarre and totally unexpected result. The number π is normally associated with circles and the numbers $1, 4, 9, \ldots$ arise in conjunction with squares and yet we somehow have a connection between all of these numbers using an infinite series.

Use the first 10 terms to approximate π. How accurate is the approximation?

13. Use the first 5 terms of the series expansion of $\exp(x)$ to approximate $\exp(0.4)$. Calculate the maximum possible error in the approximation.

14. Suppose that the first 6 terms of the series

$$1 + x + \frac{x^2}{2!} + \frac{x^3}{3!} + \cdots + \frac{x^n}{n!} + \cdots$$

are used to approximate $e^{0.3}$. What is the value of the approximation? How accurate is the approximation?

6.4 RADIUS OF CONVERGENCE

From now on we shall focus on series of functions, that is, each term in the series is a function, rather than a number. In particular, we shall consider *power series*, which are of the form

$$\sum_{k=0}^{\infty} a_k x^k = a_0 + a_1 x + a_2 x^2 + a_3 x^3 + \cdots + a_k x^k + \cdots,$$

where a_k, $k = 1, 2, 3, \ldots$ are all constants.

Consider a function f defined by a power series:

$$f(x) = a_0 + a_1 x + a_2 x^2 + a_3 x^3 + \cdots + a_k x^k + \cdots.$$

To determine where this series converges we apply the ratio test, assuming that all the coefficients are non-zero.

$$\lim_{n \to \infty} \left| \frac{a_{n+1} x^{n+1}}{a_n x^n} \right| = \lim_{n \to \infty} \left| \frac{a_{n+1} x}{a_n} \right|$$

$$= |x| \lim_{n \to \infty} \left| \frac{a_{n+1}}{a_n} \right|.$$

There are three cases to distinguish.

- Suppose that there is a positive number ℓ such that

$$\lim_{n \to \infty} \left| \frac{a_{n+1}}{a_n} \right| = \ell.$$

If we choose x so that $|x| < 1/\ell$, then

$$\lim_{n \to \infty} \left| \frac{a_{n+1} x^{n+1}}{a_n x^n} \right| = |x|\ell < 1,$$

and so the series will converge by the ratio test. The number $1/\ell$ is called the *radius of convergence* and is usually denoted by the symbol R. The series converges for all $|x| < R$ and diverges for all x such that $|x| > R$. The interval $(-R, R)$ is called the *interval of convergence*.

- If

$$\lim_{n \to \infty} \left| \frac{a_{n+1}}{a_n} \right| = 0,$$

then the series converges for every value of x and we say that the radius of convergence is infinite. We write $R = \infty$. The interval of convergence is $(-\infty, \infty)$.

- If
$$\lim_{n \to \infty} \left| \frac{a_{n+1}}{a_n} \right| = \infty,$$
then the series converges only for $x = 0$. We say that the radius of convergence is zero. There is no interval of convergence.

From this, we conclude that every power series has a radius of convergence and the series converges in a symmetric interval about the origin. There are three further points we should note:

- The series may or may not converge at the end points of the interval, that is, at the points $x = \pm R$. We shall not investigate this issue in any detail.
- It may be that $\lim_{n \to \infty} |a_{n+1}/a_n|$ does not exist. The series will still have a radius of convergence, but it will be more difficult to determine. This will not occur in any of our examples.
- We shall meet cases of power series where only even or only odd powers of x occur. In this case some of the constants a_k will be zero and the above arguments will need to be modified. This does not cause any problem and we shall deal with such cases as they arise. Such series still have a radius of convergence and there will usually be no difficulty in determining it.

EXAMPLE 6.7
We have already discussed the series
$$1 + x + x^2 + x^3 + \cdots$$
in Example 6.2 and the results we got there show that its interval of convergence is $(-1, 1)$. Let us derive the same result by applying the ratio test to the series. We have
$$\lim_{n \to \infty} \left| \frac{x^{n+1}}{x^n} \right| = \lim_{n \to \infty} |x|$$
$$= |x|$$
$$< 1,$$
as long as x is in the interval $(-1, 1)$.

We also know that
$$1 + x + x^2 + x^3 + \cdots = \frac{1}{1-x}.$$
Since we cannot put $x = 1$ in the right hand side, the series cannot converge for this point. There is no problem with any $x < 0$ and we might have expected the series to converge for all negative values of x. However, since the interval of convergence of a power series has to be a symmetric interval about the origin, and the series diverges at $x = 1$, so the interval must be $(-1, 1)$.

EXAMPLE 6.8
Determine the radius of convergence of the series
$$\sum_{n=0}^{\infty} \frac{x^{2n}}{2^n} = 1 + \frac{x^2}{2} + \frac{x^4}{2^2} + \frac{x^6}{2^3} + \cdots$$

92 SERIES AND THE EXPONENTIAL FUNCTION

Here

$$\lim_{n\to\infty}\left|\frac{u_{n+1}}{u_n}\right| = \lim_{n\to\infty}\left|\frac{x^{2(n+1)}}{2^{n+1}} \times \frac{2^n}{x^{2n}}\right|$$

$$= \lim_{n\to\infty}\left|\frac{x^2}{2}\right|$$

$$< 1,$$

as long as $|x^2| < 2$, that is, if $-\sqrt{2} < x < \sqrt{2}$. Hence the radius of convergence is $\sqrt{2}$.

EXERCISES 6.4

Find the radius of convergence and interval of convergence of each of the following power series. Ignore the problem of convergence at the end points.

1. $\displaystyle\sum_{n=1}^{\infty} \frac{2^n}{n!} x^n$

2. $\displaystyle\sum_{n=1}^{\infty} (-1)^n \frac{n^3}{3^n} x^n$

3. $\displaystyle\sum_{n=1}^{\infty} 2^n x^{2n}$

4. $\displaystyle\sum_{n=1}^{\infty} (-1)^n \frac{x^{2n}}{4^{n+1}}$

5. $\displaystyle\sum_{n=1}^{\infty} 5n^2 x^n$

6. $\displaystyle\sum_{n=1}^{\infty} (-1)^n \frac{x^{2n+1}}{(2n+1)!}$

7. $\displaystyle\sum_{n=1}^{\infty} \frac{x^{2n+1}}{3^n(2n+1)}$

8. $\displaystyle\sum_{n=1}^{\infty} n!\, x^n$

9. $\displaystyle\sum_{n=1}^{\infty} \frac{(-1)^n}{n!(n+1)!}\left(\frac{x}{2}\right)^{2n+1}$

10. $\displaystyle\sum_{n=1}^{\infty} \frac{1\cdot 4\cdot 7\cdot\ldots\cdot(3n-2)}{(3n)!} x^{3n}$

Find the interval of convergence of each of the series in Exercises 11–15. Ignore the problem of convergence at the end points.

11. $\displaystyle\sum_{n=1}^{\infty} \frac{(-1)^n}{n}(x-1)^n$

12. $\displaystyle\sum_{n=1}^{\infty} \frac{3n^2}{5^n}(x+2)^n$

13. $\displaystyle\sum_{n=1}^{\infty} \frac{(x+2)^{2n}}{2^n n!}$

14. $\displaystyle\sum_{n=1}^{\infty} (-1)^n \frac{n}{3n+1}(x+1)^{2n}$

15. $\displaystyle\sum_{n=1}^{\infty} \frac{(2n-1)3^n}{n!}(x-3)^n$

16. Verify that
$$\frac{1}{n^2} = \frac{1}{n(n+1)} + \frac{1}{n^2(n+1)}.$$

Hence, using the result of question 7 in Exercises 6.2, show that

$$\sum_{n=1}^{\infty} \frac{1}{n^2} = 1 + \sum_{n=1}^{\infty} \frac{1}{n^2(n+1)}.$$

We could compute
$$\sum_{n=1}^{\infty} \frac{1}{n^2}$$

to a given accuracy by using either one side or the other of the equation above. Which expression would be more efficient for doing this? Explain your reasons.

[This is a technique known as *improving the convergence* of a series. It has obvious practical implications.]

6.5 DIFFERENTIATION OF POWER SERIES

We now present a very important result. It tells us how to differentiate a function defined by a power series inside its radius of convergence. The result is proved using differentiation by first principles, but the proof is rather difficult and we shall not present it here. The result is as follows.

> **THEOREM 6.2 The derivative of a power series**
> *Let f be a function defined by a power series*
> $$f(x) = a_0 + a_1 x + a_2 x^2 + a_3 x^3 + \cdots + a_k x^k + \cdots$$
> *and let the radius of convergence be $R > 0$. Then the function f is differentiable everywhere in the interval $(-R, R)$ and*
> $$f'(x) = a_1 + 2a_2 x + 3a_3 x^2 + \cdots + k a_k x^{k-1} + \cdots.$$
> *The radius of convergence of the series for the derivative is also R.*

This result is not surprising, but it does tell us something very new. We know from Theorem 4.2 that the derivative of a finite sum of functions is the sum of the derivatives and the result above tells us that the same is true for an infinite sum of positive integer power functions. However, it is *not true* for all infinite sums of functions. In fact, an infinite sum of differentiable functions may not even be continuous and therefore certainly not differentiable. A simple example is the series

$$\frac{x^2}{1+x^2} + \frac{x^2}{(1+x^2)^2} + \frac{x^2}{(1+x^2)^3} + \cdots$$

This is a geometric series and for any $x \neq 0$, its sum is 1. If $x = 0$, then the sum is 0. Hence the function is not continuous at $x = 0$, and so is not differentiable there. It is, of course, not a power series.

EXAMPLE 6.9
As a first example of the use of Theorem 6.2, consider the exponential function

$$\exp(x) = 1 + x + \frac{x^2}{2!} + \frac{x^3}{3!} + \cdots$$

The above result tells us that

$$\exp'(x) = 0 + 1 + \frac{2x}{2!} + \frac{3x^2}{3!} + \cdots + \frac{nx^{n-1}}{n!} + \cdots$$
$$= 1 + x + \frac{x^2}{2!} + \frac{x^3}{3!} + \cdots$$
$$= \exp(x).$$
□

The above example exhibits a remarkable property of the exponential function, namely that the function and its derivative are identical. The only functions which have this property are the functions $k \exp x$, where $k \in \mathbb{R}$.

THEOREM 6.3 The derivative of the exponential function
For all $x \in \mathbb{R}$, we have

$$\frac{d}{dx}\exp(x) = \exp(x).$$

We can now see that the exponential function is a solution of the initial value problem $y'(x) = y(x)$, $y'(0) = 1$. We still have not solved the equation

$$p'(h) = -0.000108 p(h) - 0.013, \quad p(0) = 1013$$

of Section 6.1, but we are making progress.

EXERCISES 6.5

1. Find the radius of convergence of the series

   $$f(x) = x + 2x^2 + 3x^3 + 4x^4 + 5x^5 + 6x^6 + 7x^7 + \ldots$$

   What is the derivative of f? Justify your answer.

2. Let

   $$f(x) = a_0 + a_1 x + a_2 x^2 + a_3 x^3 + \cdots = \sum_{k=0}^{\infty} a_k x^k,$$

   for $-R < x < R$. Explain why the conclusion of Theorem 6.2 can be written in the form

   $$\frac{d}{dx}\left(\sum_{k=0}^{\infty} a_k x^k\right) = \sum_{k=1}^{\infty} k a_k x^{k-1},$$

   for $-R < x < R$. In particular, explain why the index of summation k starts from 0 in the first sum and from 1 in the second sum.

3. Explain why the functions
$$f(x) = x - \frac{x^3}{3!} + \frac{x^5}{5!} - \frac{x^7}{7!} + \cdots$$
and
$$g(x) = 1 - \frac{x^2}{2!} + \frac{x^4}{4!} - \frac{x^6}{6!} + \cdots$$
are differentiable for all $x \in \mathbb{R}$. What are the derivatives of f and g?

4. Show that the functions f and g of the previous exercise both satisfy the differential equation $y'' + y = 0$.

5. Explain why the function
$$f(x) = x - \frac{x^3}{3} + \frac{x^5}{5} - \frac{x^7}{7} + \cdots$$
is differentiable for $-1 < x < 1$. Show that f satisfies the differential equation
$$f'(x) = \frac{1}{1+x^2}.$$

6. Show that
$$\frac{1}{1-x} = 1 + x + x^2 + x^3 + \cdots, \quad -1 < x < 1.$$
Hence show that
$$\frac{1}{(1-x)^2} = 1 + 2x + 3x^2 + 4x^3 + \cdots, \quad -1 < x < 1.$$

7. The functions $\sinh x$ and $\cosh x$ are defined in terms of the exponential function by
$$\sinh x = \tfrac{1}{2}(e^x - e^{-x})$$
and
$$\cosh x = \tfrac{1}{2}(e^x + e^{-x}).$$

Show that
$$\sinh x = x + \frac{x^3}{3!} + \frac{x^5}{5!} + \frac{x^7}{7!} + \cdots$$
$$\cosh x = 1 + \frac{x^2}{2!} + \frac{x^4}{4!} + \frac{x^6}{6!} + \cdots$$

Hence show that

$$\frac{d}{dx}(\sinh x) = \cosh x$$
$$\frac{d}{dx}(\cosh x) = \sinh x.$$

The functions $\sinh x$ and $\cosh x$ are *hyperbolic functions*. We shall meet them again in Chapter 11.

6.6 THE CHAIN RULE

The exponential function satisfies a number of algebraic identities which can be used to simplify computations. One example is the identity

$$\exp(a + b) = \exp a + \exp b.$$

We shall prove this and other identities in the next section, but the proof requires a new property of the derivative, which we have not yet encountered.

In this section we shall develop a new rule for differentiation called the *chain rule*. As well as enabling us to prove identities for the exponential function, the chain rule can be used to differentiate classes of functions other than those we have already considered.

In Chapter 2, we encountered sums, multiples, products and quotients of functions. We now deal with another way of combining two functions f and g to get what is called the *composition* of these two functions. The composition is obtained by applying the two functions in succession. For example, if

$$f(x) = x^2 + 1, \quad g(x) = \frac{1}{x},$$

then

$$g(f(x)) = g(x^2 + 1) = \frac{1}{x^2 + 1}.$$

In this way we have a new function defined by the rule

$$x \mapsto \frac{1}{x^2 + 1},$$

and we denote this function by the symbol $g \circ f$, that is

$$(g \circ f)(x) = g(f(x)) = \frac{1}{x^2 + 1}.$$

> **DEFINITION 6.4 The composition of two functions**
> Let f and g be functions such that the domain of f includes the range of g. The **composition** $f \circ g$ of f and g is defined by
> $$(g \circ f)(x) = g(f(x)).$$

The process of composition is shown diagrammatically in Figure 6.5. The two function machines for f and g acting in sequence are replaced by a single machine $g \circ f$.

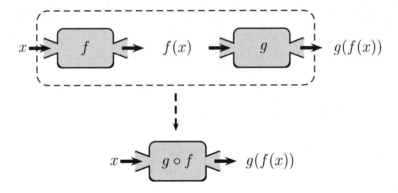

Figure 6.5: The compostion of two functions

In general $g \circ f$ and $f \circ g$ are different functions. In the above case,

$$(f \circ g)(x) = f(g(x)) = f\left(\frac{1}{x}\right) = \left(\frac{1}{x}\right)^2 + 1.$$

We also have to be careful about domains and ranges. In order for $(g \circ f)(x)$ to be well-defined, x has to be in the domain of f and $f(x)$ has to be in the domain of g.

EXAMPLE 6.10
Let $f(x) = x^2$ and let $g(x) = 1/(x-1)$. Determine the functions $f \circ g$ and $g \circ f$, together with their domains.
Solution. The function $f \circ g$ is given by

$$(f \circ g)(x) = f\left(\frac{1}{x-1}\right)$$
$$= \left(\frac{1}{x-1}\right)^2.$$

The domain of g is $\mathbb{R} \setminus \{1\}$, so this point must be excluded from the domain of $f \circ g$. For any other point $x \in \mathbb{R}$ the number $g(x) = 1/(1-x)$ will be in the domain of f, so that the domain of $f \circ g$ will be $\mathbb{R} \setminus \{1\}$.

The function $g \circ f$ is defined by

$$(g \circ f)(x) = g(x^2)$$
$$= \left(\frac{1}{x^2 - 1}\right).$$

The domain of f is $\mathbb{R}$, but if we take either of the points $x = \pm 1$, then $f(x) = x^2 = 1$, which is not in the domain of g, so that the domain of $g \circ f$ is $\mathbb{R} \setminus \{-1, 1\}$ □

We now state the chain rule for the derivative of a composite function.

THEOREM 6.4 The chain rule
If g is differentiable at x and f is differentiable at $g(x)$, then the composition $f \circ g$ is differentiable at x and

$$(f \circ g)'(x) = f'(g(x))g'(x).$$

This is a hard result to prove properly and we shall not attempt to do so. The theorem is extremely useful. Commonly, the hardest theorems to prove are the most useful in practice because, in effect, the difficulty of the calculation is transferred to the proof of the theorem.

EXAMPLE 6.11
If $f(x) = x^2 + 1$ and $g(x) = 1/x$, then

$$g'(x) = -\frac{1}{x^2}$$

and

$$f'(x) = 2x.$$

Consequently,

$$g'(f(x)) = g'(x^2 + 1)$$
$$= -\frac{1}{(x^2 + 1)^2},$$

and hence

$$(g \circ f)'(x) = g'(f(x))f'(x)$$
$$= -\frac{2x}{(x^2 + 1)^2}.$$

As an exercise, this result can be checked by using the quotient rule for $1/(x^2 + 1)$. □

There is an alternative notation for the chain rule. It is easier to understand and work with, but it lacks precision in certain cases. If we want the derivative of $g(f(x))$, then we put $u = f(x)$ and $y = g(f(x)) = g(u)$ and the rule is

$$\frac{dy}{dx} = \frac{dy}{du}\frac{du}{dx}.$$

EXAMPLE 6.12
If $f(x) = x^2 + 1$ and $g(x) = 1/x$, then

$$y = g(f(x)) = \frac{1}{x^2 + 1}.$$

Put $u = x^2 + 1$, so that

$$y = \frac{1}{u}.$$

Then

$$\frac{dy}{dx} = -\frac{1}{u^2} \times 2x$$
$$= -\frac{2x}{(x^2 + 1)^2}$$

EXERCISES 6.6

In exercises 1–6 you are given two functions f and g. In each case, find the compositions $f \circ g$ and $g \circ f$.

1. $f(x) = x^2$, $g(x) = 1 + x$
2. $f(x) = \dfrac{1}{x-3}$, $g(x) = \sqrt{x+1}$
3. $f(x) = x^2 + x$, $g(x) = \sqrt{x}$
4. $f(t) = t^2$, $g(u) = u + 3$
5. $f(t) = 2 + \dfrac{1}{t}$, $g(t) = t^2 - 4$
6. $f(x) = x^3$, $g(x) = 5$

Differentiate each of the functions in Exercises 7–10.

7. $f(x) = (1 + x^2)^2$
8. $f(x) = (1 + x^3)^4$
9. $g(t) = (t^3 + 2t^2 + 1)^4$
10. $h(\phi) = \left(\phi + \dfrac{1}{\phi}\right)^6$

6.7 PROPERTIES OF THE EXPONENTIAL FUNCTION

As we have already mentioned, the exponential function satisfies a number of algebraic identities which can be used to simplify computations. The following theorem summarises the main ones.

THEOREM 6.5

For any real numbers a and b, the exponential function satisfies the following identities.

- $\exp(a + b) = \exp(a)\exp(b)$
- $\exp(-a) = \dfrac{1}{\exp(a)}$
- $\exp(a) > 0$
- $\exp(a - b) = \dfrac{\exp(a)}{\exp(b)}$

The trick to deriving the properties of the exponential function in Therorem 6.5 is to consider the function defined on $\mathbb{R}$ by
$$f(x) = \exp(x)\exp(a + b - x),$$
where a and b are any real numbers. Using first the product rule and then the chain rule, we have

$$\begin{aligned}
\frac{d}{dx}(\exp(x)\exp(a+b-x)) \\
= \exp(x)\frac{d}{dx}(\exp(a+b-x)) + \frac{d}{dx}(\exp(x))\exp(a+b-x) \\
= \exp(x)\exp(a+b-x)\frac{d}{dx}(a+b-x) + \exp(x)\exp(a+b-x) \\
= \exp(x)\exp(a+b-x)(-1) + \exp(x)\exp(a+b-x) \\
= 0.
\end{aligned} \quad (6.13)$$

This result holds for all real numbers a, b and x. We know that if a function is constant then its derivative is zero. It is an important result that a partial converse is also true.

> **THEOREM 6.6**
> *If the derivative of a function is zero on an open interval I (possibly I = $\mathbb{R}$), then the function is constant on I.*

This result is very important and surprisingly difficult to prove and we shall not do it here. Nevertheless, we shall make use of it in equation (6.13) and in many later situations. Since f and f' are defined everywhere on $\mathbb{R}$, this equation implies that $f(x) = c$ for some constant c. Thus

$$\exp(x)\exp(a + b - x) = c.$$

This is true for all values of x, so to find c, we substitute any value of x and solve for c in terms of a and b. The obvious value to try is $x = 0$, since we know $\exp(0) = 1$. This gives

$$\exp(0)\exp(a + b) = c,$$

so that $c = \exp(a + b)$ and hence

$$\exp(x)\exp(a + b - x) = \exp(a + b). \tag{6.14}$$

Finally, let $x = a$ in this last equation to get

$$\exp(a)\exp(b) = \exp(a + b),$$

which is the first identity in Therorem 6.5.

We can also extract the second identity in Theorem 6.5 from equation (6.14). If we put $a = 0$ and $b = 0$ we find

$$\exp(x)\exp(-x) = \exp(0) = 1,$$

so that

$$\exp(-x) = \frac{1}{\exp(x)}.$$

If $x \geqslant 0$, then $\exp(x) > 0$ as can be seen from the series defining $\exp(x)$. This last equation then implies that $\exp(-x) > 0$. Thus $\exp(x) > 0$ for all x. The proof of the final identity of Theorem 6.5 is left to you as an exercise.

The properties of the exponential function in Theorem 6.5 enable us to develop an alternative way of looking at this function. We first define a number e by letting $e = \exp(1)$. Calculations with the power series give $e \doteq 2.71828$. The number e is an irrational number and it plays a key role in the applications of calculus to the real world. Once we have defined e, we have the following.

$$\exp(2) = \exp(1 + 1) = \exp(1)\exp(1) = e^2$$

and similarly $\exp(3) = e^3$, $\exp(4) = e^4$,...
Also

$$\exp(-2) = \frac{1}{\exp(2)} = \frac{1}{e^2} = e^{-2}$$

and similarly $\exp(-3) = e^{-3}$, $\exp(-4) = e^{-4}$,... These results, together with the result $\exp(0) = 1$, show that

$$\exp(n) = e^n$$

for all integers n.

How would we calculate the square root of the number e? We have as yet given no rule for calculating the square root of any number which is not an exact square, but we can now give a procedure for calculating $\sqrt{e}$. We have

$$e = \exp 1 = \exp(\tfrac{1}{2} + \tfrac{1}{2}) = \left(\exp(\tfrac{1}{2})\right)^2,$$

so that

$$\exp(\tfrac{1}{2}) = \sqrt{e}.$$

To calculate $\exp(\tfrac{1}{2})$ and hence $\sqrt{e}$, we use the power series for the exponential function. We can similarly deal with any rational power $r = p/q$, where p and q are integers. We have

$$e^p = \exp(p) = \exp\underbrace{\left(\frac{p}{q} + \cdots + \frac{p}{q}\right)}_{q \text{ terms}}$$

$$= \left(\exp\left(\frac{p}{q}\right)\right)^q$$

$$= (\exp(r))^q.$$

Hence

$$e^r = e^{p/q} = (e^p)^{1/q} = \exp(r),$$

so that we can calculate e^r by using the series for $\exp(r)$. Finally, when α is an irrational number, we simply *define* e^α by letting $e^\alpha = \exp(\alpha)$. We then have

$$e^x = \exp x$$

for all $x \in \mathbb{R}$.

EXAMPLE 6.13
Calculate $\sqrt{e}$ to 4 decimal places.
Solution. We have

$$\exp(x) = 1 + x + \frac{x^2}{2!} + \frac{x^3}{3!} + \cdots$$

The error E_n satisfies

$$E_n \leqslant \frac{x^n}{(n-1)!(n-x)}.$$

If we put $x = 1/2$, then we find $E_7 < 0.00003$, so that

$$\sqrt{e} \simeq 1 + \frac{1}{2} + \cdots + \frac{1}{2^6 6!} \doteq 1.64872$$

to 4 decimal places.

EXAMPLE 6.14
Find the derivatives of $\exp(ax)$ and $\exp(-ax)$.
Solution. We use the chain rule, together with the result $\exp'(x) = \exp(x)$.

$$\frac{d}{dx}(\exp(ax)) = \exp(ax)\frac{d}{dx}ax = a\exp(ax).$$

Similarly
$$\frac{d}{dx}(\exp(-ax)) = \exp(-ax)\frac{d}{dx}(-ax) = -a\exp(-ax).$$
Another way of writing these results is
$$\frac{d}{dx}e^{ax} = ae^{ax}, \qquad (6.15)$$
$$\frac{d}{dx}e^{-ax} = -ae^{-ax}. \qquad (6.16)$$

EXERCISES 6.7

Differentiate the functions given in Exercises 1–8.

1. $y = e^{-3x}$
2. $y = e^{-x^2}$
3. $f(t) = \frac{1}{2}(e^t + e^{-t})$
4. $f(x) = (1 - e^{3x})^4$
5. $f(x) = \dfrac{e^x - 1}{e^x + 1}$
6. $y = x^3 e^{2x}$
7. $g(x) = 2xe^{x^2+x+1}$
8. $g(x) = \dfrac{e^x}{1+x}$

9. Using the series definition of the exponential function, find an approximation for $e^{0.2}$ accurate to 3 decimal places. This means that your maximum error must be less than 0.0005.

10. Find an approximate value of $1/e$ to 3 decimal place accuracy, using the series expansion for the exponential function.

11. How many terms of the exponential series would we need in order to find e^5 accurate to 0.0000000001, that is, accurate to 10^{-10}?

12. Find an example of a function f for which $f'(x) = 0$ for all x in the domain of f, but for which $f(x)$ is not a constant. Can you reconcile this result with the statement of Theorem 6.6?

6.8 SOLUTION OF THE AIR PRESSURE PROBLEM

In Section 6.1, we found the differential equation for the air pressure problem to be
$$p'(h) = -0.000108p(h) - 0.013, \quad p(0) = 1013. \qquad (6.17)$$
The constants are awkward to deal with, so let us replace them by symbols for ease of working. We write
$$p'(h) = -ap(h) + b, \quad p(0) = c,$$
with $a = 0.000108$, $b = -0.013$ and $c = 1013$. Equations of this form are applicable to problems other than the air pressure problem, so let us emphasize this generality by writing $y = f(x)$, where f has no particular physical interpretation. The differential equation to be solved is
$$y' + ay = b, \quad y(0) = c. \qquad (6.18)$$

We could solve this equation by a power series method, but since we have now developed an impressive amount of mathematics in defining the exponential function, let us try to make use of some of it.

SOLUTION OF THE AIR PRESSURE PROBLEM

We solve equation (6.18) by working through several steps. First we solve the case when $b = 0$, that is, we solve the equation

$$y' + ay = 0. \qquad (6.19)$$

At this stage we will ignore the initial condition $y(0) = c$. To solve the equation we make use of the result (6.16), which shows that $y = e^{-ax}$ is a solution of the differential equation. Sometimes equations have more than one solution, so we have to ask whether the solution $y = e^{-ax}$ is the only solution of equation (6.19). Suppose that $y_1(x)$ is some other solution, so that $y_1'(x) = -ay_1(x)$. Then, by the quotient rule,

$$\frac{d}{dx}\left(\frac{y_1(x)}{e^{-ax}}\right) = \frac{1}{e^{-2ax}}\left(e^{-ax}y_1'(x) - y_1(x)\frac{d}{dx}(e^{-ax})\right)$$

$$= \frac{1}{e^{-2ax}}\left(e^{-ax}(-ay_1(x)) - y_1(x)(-a)(e^{-ax})\right)$$

$$= 0.$$

This means that

$$\frac{y_1(x)}{e^{-ax}} = A,$$

for some constant A and so

$$y_1(x) = Ae^{-ax}.$$

This result shows that *every* solution of $y' + ay = 0$ must have this form. We have reached an important result which is worth some discussion. We call

$$y = Ae^{-ax}$$

the *general solution* of the equation $y' + ay = 0$, since every solution has this form. If we know the initial condition $y(0) = y_0$, then it follows that $A = y_0$, so that

$$y = y_0 e^{-ax}$$

is a solution of the initial value problem $y' + ay = 0$, $y(0) = y_0$. We call $y = y_0 e^{-ax}$ a *particular solution* of $y' + ay = 0$. In summary we have the following result.

The initial value problem

$$y'(x) + ay(x) = 0, \quad y(0) = y_0 \qquad (6.20)$$

has the solution

$$y = y_0 e^{-ax}.$$

Next we move on to the more general equation

$$y' + ay = b.$$

The strategy here is to replace the constant A by a function $K(x)$. The idea behind this is as follows:

1. $y = Ae^{-ax}$ satisfies $y' + ay = 0$.

2. If we replace the constant A by a function $K(x)$, we may be able to choose $K(x)$ so that the right hand side comes out to b rather than zero.

This is an instance of a general procedure for solving differential equations we will encounter again later. It is known as the method of *variation of parameters* or *variation of constants*. The idea is to solve a simpler problem and then replace the constants by functions in a trial solution for the more complex problem. If we substitute
$$y = K(x)e^{-ax}$$
into the equation
$$y' + ay = b,$$
we get
$$K'(x)e^{-ax} - aK(x)e^{-ax} + aK(x)e^{-ax} = b,$$
so that
$$K'(x)e^{-ax} = b$$
and hence
$$K'(x) = be^{ax}.$$
We are now looking for a function $K(x)$ whose derivative is be^{ax}. It is not difficult to verify, using the chain rule, that
$$K(x) = \frac{b}{a}e^{ax} + B$$
is such a function, where B is a constant. Thus the solution we are seeking has the form
$$y = K(x)e^{-ax} = \left(\frac{b}{a}e^{ax} + B\right)e^{-ax}$$
$$= \frac{b}{a} + Be^{-ax}.$$
This is easily checked to be a solution of $y' + ay = b$ for any value of the constant A. If we now make use of the initial condition $y(0) = c$, we have $A = c - b/a$ and so
$$y = \frac{b}{a} + \left(c - \frac{b}{a}\right)e^{-ax}.$$
This is another result worth summarising:

> The initial value problem
> $$y' + ay = b, \quad y(0) = y_0 \qquad (6.21)$$
> has the solution
> $$y = \frac{b}{a} + \left(y_0 - \frac{b}{a}\right)e^{-ax}.$$

The air pressure equation (6.17) has the same form as equation (6.21). Using the values $a = 0.000108$, $b = -0.013$, $p_0 = c = 1013$ in equation (6.17) gives
$$p(h) = -120 + 1133e^{-0.000108h}$$
as the solution to the air pressure problem. We can check that this will reproduce the results of Table 6.1. In computing the exponential terms we *could* use the power series expansions, but almost any calculator will have the calculation hardwired into it, so that we can evaluate the exponentials at

the touch of a button. (But remember, the designers of the calculator needed to have access to much of the theory of this chapter.)

The unifying thread of this chapter has been the problem of finding an equation for the variation of air pressure with respect to height and we have developed an impressive amount of mathematics to reach the solution. The big payoff, however, is not that we can now solve the air pressure problem. Rather, it is that the mathematics we have developed is useful and applicable in a wide variety of seemingly unrelated problems. One of these is given in the example below. Others appear in Exercises 6.8.

EXAMPLE 6.15

A tank initially contains 100 litres of water in which 5 kilograms of salt is dissolved. Pure water is pumped into the tank at a rate of 2 litres per second. The mixture in the tank is pumped out at the same rate. Set up and solve the differential equation for the amount of salt x in the tank at time t.

Solution. Let x be the amount of salt in the tank at time t. Then the initial condition is $x(0) = 5$ and the amount of salt per litre is $x/100$. Let Δx be change in the amount of salt in the tank during the time interval $[t, t + \Delta t]$, where Δt is assumed to be small. We have

$$\Delta x \approx -\frac{x}{100} \times 2 \times \Delta t$$

and this approximation becomes increasingly accurate as we let Δt approach zero. We rearrange this equation in the form

$$\frac{\Delta x}{\Delta t} \approx -\frac{2x}{100}.$$

Using the Leibniz approach to differentiation (page 46), we let Δt approach zero and the equation becomes

$$\frac{dx}{dt} = -\frac{x}{50}.$$

This is of the same form as equation (6.20) with $a = 1/50$. The solution is

$$x(t) = 5e^{-t/50}.$$

EXERCISES 6.8

The exercises in this section can (and should) be done without the use of logarithms, which we consider in Chapter 9.

A simple model of population growth assumes that the population P grows at a rate which is proportional to P. That is

$$\frac{dP}{dt} = kP,$$

where k is a positive constant. This is often called the *exponential* or *natural growth* equation. It applies if birth and death rates are constant.

As well as representing population growth, the above differential equation (possibly with $k < 0$) has many other applications, including compound interest, radioactive decay and drug elimination from the bloodstream.

Use the exponential growth equation in exercises 1–4.

1. A city had a population of 25 000 in 1980 and a population of 30 000 in 1990. What will the population be in the year 2000?

2. The number of bacteria in a certain culture increases at a rate proportional to the number present. Suppose there are 1000 bacteria present initially and 1600 after 1 hour. How many will be present after 3 hours?

3. The drug sodium pentobarbitol is used for anesthetising dogs. The dog is anesthetised as long as its bloodstream contains at least 45 mg of the drug per kilogram of the dog's body weight. Suppose that sodium pentobarbitol is eliminated from the dog's bloodstream at a rate proportional to the amount of the drug present, and that half the drug is eliminated after 5 hours. How much of the drug should be administered in order to anesthetise a 25 kg dog for 2 hours?

4. A curve passes through the point $(0, 4)$ and has the property that the slope of the curve at any point P is three times the value of the y coordinate of P. Find the equation of the curve.

5. Let T be the temperature of a body. *Newton's law of cooling* states that the time rate of change of the temperature $T(t)$ of a body is proportional to the difference between T and the temperature T_0 of the surrounding medium. Show that

$$\frac{dT}{dt} = -k(T - T_0),$$

where $k > 0$ is a constant.

Solve this equation in the following two ways:

(a) Use equation (6.21) in the text.

(b) Put $U = T - T_0$ and show that the equation becomes

$$\frac{dU}{dt} = -kU.$$

Solve this equation and replace U by $T - T_0$ in the solution. What is the physical interpretation of U?

6. Solve the initial value problem $y' = -ay + b$, $y(0) = c$ in the following way.

(a) Write the differential equation as

$$y' = -a\left(y - \frac{b}{a}\right).$$

(b) Put $u = y - b/a$, solve the resulting initial value problem and replace u by $y - b/a$ in the solution.

7. A cake is removed from an oven at a temperature of 150°C and left to cool. After 30 min, the temperature of the cake is 90°C. What will the temperature be after 2 hours? Assume that the room temperature is 25°C.

8. A tank initially contains 100 litres of pure water. Water containing salt at a concentration of 0.1 kg/litre is pumped into the tank at a rate of 5 litre/min. The mixture in the tank is pumped out at the same rate. Show that the differential equation for the amount of salt x in the tank at time t is
$$\frac{dx}{dt} = 0.5 - \frac{x}{20}.$$
Hence find the amount of salt in the tank after 10 min.

9. A lake has a volume of 1 billion cubic metres of water, with an initial pollutant concentration of 0.25%. The average daily inflow of water into the lake is 1 million cubic metres of water, with an average pollutant concentration of 0.05%. The average daily outflow of well mixed water from the lake is equal to the average inflow. What will be the concentration of pollutant after 5 years?

10. Using *Mathematica*, check that the solution to the air pressure problem
$$y = -120 + 1133e^{-0.000108h}$$
obtained above will closely reproduce the values in Table 6.1.

The atmosphere only extends a finite distance above the surface of the earth. Consequently, we would expect the pressure to approach zero as the height h increases indefinitely. Suggest reasons why this does not happen for the above solution.

CHAPTER 7

TRIGONOMETRIC FUNCTIONS

In this chapter we consider the vibrating cables of the Anzac Bridge. A full analysis of the problem would take us well beyond the mathematics in this book, but we can still make progress by making some assumptions and simplifications.

7.1 VIBRATING STRINGS AND CABLES

The problem to be considered is the vibration of a cable or string with fixed end points. On a large scale this could apply to the Anzac Bridge, while on a small scale it could apply to a musical instrument, such as a violin or guitar. The numerical information we want to derive is the frequency at which the cable vibrates. The frequency we are considering here is the frequency of vibration when there are no outside influences. This is called the *natural frequency*. If some outside influence tries to make a system vibrate with its natural frequency, then the system will readily do so. To make it vibrate at a different frequency is more difficult, a problem we will consider in detail in Chapter 8.

Let us consider a vibrating string or cable. The mass is spread out along the string. If the total mass is m and the string has length ℓ, then its mass per unit length is $\rho = m/\ell$. To analyse the motion of the string, we have to consider each part of it, but this requires mathematics beyond that which we will develop in this book. This is a task for a later subject. Instead, we shall make a simple approximation and assume that the mass of the string is concentrated in a single body at the center of the string (Figure 7.1). The string vibrates because it is under tension and this tension exerts a force on the body. We denote the tension in each half of the string by T_1 and T_2 respectively and we let x be the displacement of the centre of the string from the equilibrium position.

The tension in each half of the string can be split into two parts, a horizontal component and a vertical component. Since the body does not move horizontally, the two horizontal components T_1^h and T_2^h must be equal and opposite to each other and we can ignore them as far as the vertical motion is concerned. It follows that the tensions T_1 and T_2 are equal and hence that the two vertical components are equal. Let T be the common value of T_1 and T_2 and let T_v be the value of the vertical component. The two vertical components add up to give a total force of $2T_v$ on the body. Newton's second law of motion implies that

$$m\frac{d^2x}{dt^2} = -2T_v. \tag{7.1}$$

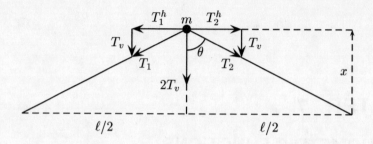

Figure 7.1: A vibrating string or cable

Using similar triangles in Figure 7.1, we have

$$\frac{T_v}{T} = \frac{x}{\sqrt{x^2 + (\ell/2)^2}}.$$

If x is small, then

$$\sqrt{x^2 + (\ell/2)^2} \approx \frac{\ell}{2},$$

so for small x,

$$\frac{T_v}{T} \approx \frac{x}{\ell/2},$$

that is,

$$T_v \approx \frac{2Tx}{\ell}.$$

We shall assume that the error in this approximation is negligible for small x and write

$$T_v = \frac{2Tx}{\ell}.$$

Using this result in equation (7.1) gives us

$$m\frac{d^2x}{dt^2} = -\frac{4Tx}{\ell},$$

or

$$\frac{d^2x}{dt^2} = -\frac{4Tx}{m\ell}$$

$$= -\frac{4Tx}{\ell^2\rho}.$$

We shall write this as
$$\frac{d^2x}{dt^2} + \omega^2 x = 0,$$
where
$$\omega = \frac{2}{\ell}\sqrt{\frac{T}{\rho}}.$$

We shall assume that the motion commences at $t = 0$ with the mass displaced by x_0 and with velocity x_1. Thus $x(0) = x_0$ and $x'(0) = x_1$, which leads to the initial value problem
$$\frac{d^2x}{dt^2} + \omega^2 x = 0,\ x(0) = x_0,\ x'(0) = x_1.$$

The differential equation in this problem is known as a *second order differential equation*, since the derivative appearing in the equation is of second order. We will discuss solutions of this equation in the next section.

7.2 TRIGONOMETRIC FUNCTIONS

In the previous section, we arrived at the initial value problem
$$\frac{d^2x}{dt^2} + \omega^2 x = 0,\ x(0) = x_0,\ x'(0) = x_1. \tag{7.2}$$

To solve this equation, we use the method of infinite series as developed in Chapter 6. Polynomial functions will fail for the same reasons that they failed in Chapter 6. Let us use, as a trial solution, a function defined by a power series:
$$x(t) = \sum_{k=0}^{\infty} a_k t^k. \tag{7.3}$$

The problem is to determine the coefficients a_k. On differentiating twice, using Theorem 6.2, we have
$$\frac{d^2x}{dt^2} = \sum_{k=2}^{\infty} k(k-1) a_k t^{k-2} \tag{7.4}$$

You may be puzzled by the fact that the sum now goes from $k = 2$ instead of from $k = 0$. The reason is easy to see if we first expand the first few terms of the sum (7.3) to get
$$x(t) = a_0 + a_1 t + a_2 t^2 + a_3 t^3 + a_4 t^4 + \cdots,$$
and then take the second derivative
$$x''(t) = 2a_2 + 3.2 a_3 t + 4.3 a_4 t^2 + \cdots = \sum_{k=2}^{\infty} k(k-1) a_k t^{k-2}.$$

Since the derivative of a constant is zero, one term drops out for each differentiation. We substitute equations (7.4) and (7.3) into equation (7.2) to get
$$\sum_{k=2}^{\infty} k(k-1) a_k t^{k-2} + \omega^2 \sum_{k=0}^{\infty} a_k t^k = 0,$$

that is, replacing k by $k+2$ in the first sum,

$$\sum_{k=0}^{\infty}(k+2)(k+1)a_{k+2}t^k + \omega^2 \sum_{k=0}^{\infty} a_k t^k = 0.$$

Hence

$$\sum_{k=0}^{\infty} \left((k+2)(k+1)a_{k+2}t^k + \omega^2 a_k t^k\right) = 0,$$

that is,

$$\sum_{k=0}^{\infty} \left((k+2)(k+1)a_{k+2} + \omega^2 a_k\right)t^k = 0.$$

If we equate coefficients of t^k on both sides of the equation for $k = 1, 2, 3, \ldots$, we find

$$(k+2)(k+1))a_{k+2} + \omega^2 a_k = 0, \quad k = 1, 2, 3, \ldots$$

which shows that

$$a_{k+2} = -\frac{\omega^2 a_k}{(k+2)(k+1)}. \tag{7.5}$$

Equation (7.5) is called a *recurrence relation*; it gives the coefficient a_{k+2} in terms of a_k. Since equation (7.5) relates each coefficient to the second one before it, the even-numbered coefficients $(a_0, a_2, a_4, a_6, \ldots)$ and the odd-numbered coefficients $(a_1, a_3, a_5, \ldots)$ will be determined separately. For the even-numbered coefficients, we have

$$a_2 = -\frac{a_0 \omega^2}{2.1} = -\frac{a_0 \omega^2}{2!}, \quad a_4 = -\frac{a_2 \omega^2}{4.3} = \frac{a_0 \omega^4}{4!}, \quad a_6 = -\frac{a_4 \omega^2}{6.5} = -\frac{a_0 \omega^6}{6!},$$

and in general

$$a_{2k} = \frac{(-1)^k}{(2k)!} \omega^{2k} a_0, \quad k = 1, 2, 3, \ldots \tag{7.6}$$

Similarly, for the odd-numbered coefficients, we have

$$a_3 = -\frac{a_1 \omega^2}{3.2} = -\frac{a_1 \omega^2}{3!}, \quad a_5 = -\frac{a_3 \omega^2}{5.4} = \frac{a_1 \omega^4}{5!}, \quad a_7 = -\frac{a_5 \omega^2}{7.6} = -\frac{a_1 \omega^6}{7!},$$

and in general

$$a_{2k+1} = \frac{(-1)^k}{(2k+1)!} \omega^{2k} a_1, \quad k = 1, 2, 3, \ldots \tag{7.7}$$

Substitution of equations (7.6) and (7.7) into equation (7.3) gives

$$x(t) = a_0 \left(1 - \frac{(\omega t)^2}{2!} + \frac{(\omega t)^4}{4!} - \cdots \right)$$

$$+ \frac{a_1}{\omega}\left(\omega t - \frac{(\omega t)^3}{3!} + \frac{(\omega t)^5}{5!} + \cdots\right) \tag{7.8}$$

The two series in parentheses are extremely important. They are functions which you may have already met, although the connection will not be obvious at this stage. We define the *sine* and *cosine* functions as follows:

DEFINITION 7.1 The sine and cosine functions
For all $t \in \mathbb{R}$, we define

$$\sin t = t - \frac{t^3}{3!} + \frac{t^5}{5!} - \frac{t^7}{7!} + \cdots = \sum_{k=0}^{\infty} \frac{(-1)^k}{(2k+1)!} t^{2k+1} \qquad (7.9)$$

$$\cos t = 1 - \frac{t^2}{2!} + \frac{t^4}{4!} - \frac{t^6}{6!} + \cdots = \sum_{k=0}^{\infty} \frac{(-1)^k}{(2k)!} t^{2k}. \qquad (7.10)$$

Of course, we have to verify that the series in Definition 7.1 actually converge. We can do this by means of the ratio test. The general term of the sine series is

$$a_k = \frac{(-1)^k}{(2k+1)!} t^{2k+1},$$

so that

$$\lim_{k \to \infty} \left| \frac{a_k}{a_{k-1}} \right| = \lim_{k \to \infty} \left| \frac{(-1)^k}{(2k+1)!} t^{2k+1} \bigg/ \frac{(-1)^{k-1}}{(2k-1)!} t^{2k-1} \right|$$

$$= \lim_{k \to \infty} \frac{|t^2|}{(2k+1)(2k-1)}$$

$$= 0.$$

Since this limit is less than 1 for every $t \in \mathbb{R}$, the series converges by the ratio test and the radius of convergence is infinite. Thus $\sin t$ has domain $\mathbb{R}$. A similar calculation shows that the series for $\cos t$ converges for all $t \in \mathbb{R}$ and so the domain of $\cos t$ is $\mathbb{R}$.

Using Definition 7.1, equation (7.8) can now be written

$$x(t) = a_0 \cos \omega t + \frac{a_1}{\omega} \sin \omega t. \qquad (7.11)$$

The arbitrary constants a_0 and a_1 can be found using the initial conditions. Finding a_0 is easy. We put $t = 0$ in equation (7.8) and find $x(0) = a_0$. To find a_1, we differentiate equation (7.8) with respect to t and then put $t = 0$. This procedure is justified by Theorem 6.2. We find

$$x'(t) = a_0\left(-\omega^2 t + \frac{t^3 \omega^4}{3!} - \frac{t^5 \omega^6}{5!} + \cdots\right) + \frac{a_1}{\omega}\left(\omega - \frac{\omega^3 t^2}{3!} + \frac{\omega^5 t^4}{4!} + \cdots\right).$$

Putting $t = 0$ gives $x'(0) = a_1$, so that $a_1 = x_1$. The final solution to the initial value problem (7.2) is thus

$$x(t) = x_0 \cos \omega t + \frac{x_1}{\omega} \sin \omega t. \qquad (7.12)$$

EXERCISES 7.2

1. Use the ratio test to show that the series (7.10) converges for all real numbers t.

2. The ends of a string under tension are fixed at two points A and B. The distance AB is 2 m, the mass per unit length of the string is 0.1 kg and the tension in the string is 2 N. The centre of the string is displaced a distance of 5 cm from its equilibrium position and then released. Find the value of the constant ω in equation (7.12) and use this equation to find the displacement of the centre of the cable after 2 seconds.

7.3 MORE ON THE SINE AND COSINE FUNCTIONS

Definition 7.1 provides the basis for all calculations with $\sin t$ and $\cos t$. These functions have been introduced because they provide solutions for the differential equation (7.2). This differential equation arises in all oscillation problems. The power series definitions enable function values to be calculated to any desired accuracy and so the functions are well-defined. We shall later relate these trigonometric functions to the properties of triangles. These properties are used to define trigonometric functions in elementary mathematics, but it should be noted that the definitions in this form do not enable function values to be calculated to arbitrary accuracy.

The series for $\sin x$ and $\cos x$ each have the property that successive terms have opposite signs for any value of t. Such a series is called an *alternating series*. There is a simple rule for calculating the error term for such a series, as long as the terms are decreasing in magnitude. The error, in absolute value, is less than the first term omitted. In the case at hand the terms of either series may initially increase for some values of t, but as long as we take enough terms, they will eventually start to decrease. For example, if $t = 4$ in the sine series, we get

$$\sin 4 = 4 - \frac{4^3}{3!} + \frac{4^5}{5!} - \frac{4^7}{7!} + \cdots$$
$$= 4 - 10\tfrac{2}{3} + 8\tfrac{8}{15} - 3\tfrac{301}{1264} + \cdots$$

If the value of t is large, there may be many increasing terms before a decrease begins. To see that the terms will always start to decrease eventually, we note that each term is obtained from the previous one by multiplying it by an expression of the form

$$\frac{-t^2}{(2k+1)(2k+2)}.$$

For large enough k the absolute value of this term will be less than 1. Thus to calculate $\cos t$, we use the following rule:
choose k so that

$$\frac{t^2}{(2k+1)(2k+2)} < 1$$

and then calculate the terms

$$\frac{t^{2k+2}}{(2k+2)!}, \frac{t^{2k+4}}{(2k+4)!}, \ldots$$

until we find one whose magnitude is less than the required error. Suppose this term is

$$\frac{t^{2n}}{(2n)!}.$$

Then

$$\cos t \approx 1 - \frac{t^2}{2!} + \frac{t^4}{4!} - \cdots + \frac{(-1)^n t^{2n-2}}{(2n-2)!},$$

with an error of less than
$$\frac{t^{2n}}{(2n)!}.$$

EXAMPLE 7.1
Suppose we wish to evaluate $\cos(0.2)$ with an error of less than 0.0001. Since
$$\frac{(0.2)^2}{2!} < 1,$$
all terms after the first are decreasing and we calculate successive terms. We find
$$\frac{(0.2)^4}{4!} \doteq 0.00007.$$
Thus we omit all terms after the second to get
$$\cos 0.2 \approx 1 - \frac{(0.2)^2}{2!} = 0.98,$$
with an error of less than 0.00007. In terms of inequalities we have
$$0.97993 < \cos 0.2 < 0.98007.$$
It is common to write results such as this in the form
$$\cos 0.2 = 0.98000 \pm 0.00007.$$

EXAMPLE 7.2
To evaluate $\cos 4$ with an error of less than 0.0001, we first look at terms of the form
$$\frac{4^2}{(2n+1)(2n+2)}.$$
This gives
$$n = 1: \quad \frac{16}{3 \times 4} > 1$$
$$n = 2: \quad \frac{16}{4 \times 5} < 1$$
This means we have to evaluate the terms
$$\frac{4^6}{6!}, \frac{4^8}{8!}, \ldots$$
until one of them is less than the allowable error. We find
$$\frac{4^{16}}{16!} \doteq 0.0002, \quad \frac{4^{18}}{18!} \doteq 0.00001,$$
so that
$$\cos 4 \approx 1 - \frac{4^2}{2!} + \frac{4^4}{4!} - \cdots + \frac{4^{16}}{16!} \doteq -0.6536,$$
with an error of less than 0.0001. □

These calculations can get very lengthy as the argument of the function gets larger, but as we shall soon see there are properties of the functions which shorten them considerably. These properties have to be deduced from the power series definitions or from the original differential equation (7.2).

Some properties are easily deduced from the power series definitions. If we put $t = 0$ in the series for $\sin t$, we find $\sin 0 = 0$. Similarly, $\cos 0 = 1$. If we replace x by $-x$ in the sine series, we see that $\sin(-x) = -\sin x$, and similarly, $\cos(-x) = \cos x$. There is, however, a fundamental property of the sine and cosine functions which is not at all obvious from either the power series or the differential equation and that is their *periodicity*. This means that there is a particular number T such that $\cos(t + T) = \cos t$ and $\sin(t + T) = \sin t$ for all values of t. We shall need some preliminary results before we can demonstrate this.

The first result we need is contained in Definition 7.1. If we use the rule for differentiation of a power series to differentiate equation (7.9) with respect to t, we find

$$\frac{d}{dt}(\sin t) = 1 - \frac{t^2}{2!} + \frac{t^4}{4!} - \frac{t^6}{6!} + \cdots,$$

while if we differentiate equation (7.10) with respect to t, we find

$$\frac{d}{dt}(\cos t) = -\left(t - \frac{t^3}{3!} + \frac{t^5}{5!} - \frac{t^7}{7!} + \cdots\right).$$

By comparing these equations with Definition 7.1, we reach the following important theorem.

THEOREM 7.1 Derivatives of the sine and cosine functions
For all $t \in \mathbb{R}$, we have

$$\frac{d}{dt}(\sin t) = \cos t$$
$$\frac{d}{dt}(\cos t) = -\sin t.$$

The next theorem states two important properties of the sine and cosine functions.

THEOREM 7.2 Identities for the sine and cosine functions
For all real numbers a and b we have

$$\sin(a + b) = \sin a \cos b + \cos a \sin b$$
$$\cos(a + b) = \cos a \cos b - \sin a \sin b.$$

This theorem can be proved by a similar method to that used to show $\exp(a + b) = \exp a \exp b$. We begin with the function

$$f(t) = \sin t \cos(a + b - t) + \cos t \sin(a + b - t).$$

Then, using the product rule and the chain rule,

$$\begin{aligned} f'(t) &= \cos t \cos(a + b - t) + \sin t \sin(a + b - t) \\ &\quad - \sin t \sin(a + b - t) - \cos t \cos(a + b - t) \\ &= 0. \end{aligned}$$

Since $f'(t) = 0$, we must have $f(t) = c$ for some constant c. Hence

$$c = \sin t \cos(a + b - t) + \cos t \sin(a + b - t).$$

If we put $t = 0$ we get $c = \sin(a + b)$, where we have used the results $\sin 0 = 0$, $\cos 0 = 1$. Putting $t = a$ then gives the first identity in Theorem 7.2. The proof of the second one is similar. If we replace b by $-b$ in this theorem and use the results $\cos(-t) = \cos t$, $\sin(-t) = -\sin t$, we deduce the following result.

> **THEOREM 7.3 Identities for the sine and cosine functions**
> *For all real numbers a and b we have*
>
> $$\sin(a - b) = \sin a \cos b - \cos a \sin b$$
> $$\cos(a - b) = \cos a \cos b + \sin a \sin b.$$

If we put $a = b = t$ in the second identity in Theorem 7.3, we get

$$\cos(t - t) = \cos t \cos t + \sin t \sin t.$$

Hence
$$\cos^2 t + \sin^2 t = 1.$$

Finally, from Theorem 7.2 we can get the following double angle formulas by putting $a = b$.

$$\sin 2a = 2 \sin a \cos a$$
$$\cos 2a = \cos^2 a - \sin^2 a.$$

using the result $\cos^2 a + \sin^2 a = 1$ in this last equation, enables us to deduce that

$$\cos 2a = 2 \cos^2 a - 1$$

and

$$\cos 2a = 1 - 2 \sin^2 a.$$

All of these results have been obtained purely from the power series definitions of the functions.

EXERCISES 7.3

Prove the trigonometric identities in Exercises 1–4.

1. $(\sin \theta + \cos \theta)^2 + (\sin \theta - \cos \theta)^2 = 2$
2. $\dfrac{1 - \sin x}{\cos x} = \dfrac{\cos x}{1 + \sin x}$
3. $\dfrac{1}{1 - \sin \theta} + \dfrac{1}{1 + \sin \theta} = \dfrac{2}{\cos^2 \theta}$
4. $\dfrac{\sin^3 x + \cos^3 x}{\sin x + \cos x} = 1 - \sin x \cos x$

In Exercises 5–12, find the derivative of the given function.

5. $f(x) = \sin^2 x$

6. $f(t) = \sin 2t \cos 3t$

7. $y = x^2 \cos x$

8. $g(\phi) = e^{2\phi} \cos \phi$

9. $f(x) = \dfrac{\sin x}{1 - \cos x}$

10. $f(x) = (1 + \sin^2 x)^3$

11. $f(x) = \cos x - \sqrt{3} \sin x$

12. $f(x) = \begin{cases} \sin x, & x \neq 0 \\ 1, & x = 0. \end{cases}$

13. Let f and g be differentiable functions such that $f'(x) = -g(x)$ and $g'(x) = f(x)$ for all x.

    (a) Prove that $f^2(x) + g^2(x) = C$, a constant

    (b) Suppose there is a number a such that $f(a) = 1$ and $g(a) = 0$. Find the value of C.

    (c) Give an example of a pair of functions with these properties.

14. Show that $y = -\frac{1}{4} x \cos 2x$ satisfies the differential equation
$$y'' + 4y = \sin 2x.$$

15. Show that $y = -\frac{1}{3} e^x \cos x$ satisfies the differential equation
$$y'' - 2y' - y = e^x \cos x.$$

16. Show that $y = Ae^x + Be^{-x} - e^x \sin e^{-x}$ satisfies the differential equation
$$y'' - y = e^{-x} \sin e^{-x} + \cos e^{-x},$$
where A and B are arbitrary constants.

17. A simple pendulum consists of mass m suspended from a fixed point by a light inelastic string, as in the diagram.

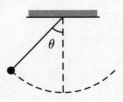

The angular displacement $\theta(t)$ of the mass at time t is given by
$$\theta(t) = A \cos(\omega t + \alpha),$$
where A, ω and α are constants.

(a) Show that θ satisfies the differential equation $\theta'' + \omega^2 \theta = 0$.

(b) Show that θ can be written in the form

$$\theta(t) = a\cos\omega t + b\sin\omega t,$$

where a and b are constants.

7.4 TRIANGLES, CIRCLES AND THE NUMBER π

You will already have met the sine and cosine functions in connection with triangles. The functions we have defined in terms of power series are the same functions, but we have not yet shown this. Notice, that while the triangle definitions may look simpler than the power series definitions, they do not give any systematic method for calculating function values. These functions can also be defined by using the unit circle, but this is just a generalisation of the triangle definitions. The number π arises naturally in connection with trigonometric functions and here we can also make connections with the power series definitions.

7.4.1 Trigonometric functions defined by triangles

In this section we connect the series definitions for the trigonometric functions with the more familiar definitions in terms of the ratios of various sides of a right-angled triangle.

Suppose we have a particle moving around a circle of radius r such that the velocity has constant magnitude $v > 0$. At each instant the direction of the velocity is tangential to the circle (Figure 7.2). The velocity can be resolved into a component v_x parallel to the x axis and a component v_y parallel to the y axis. We can get the *magnitudes* of v_x and v_y by using the fact that triangles OBA and AQP are similar, but we have to be careful about the *sign* of these quantities. Let's look at the magnitudes first.

In triangles OBA and AQP, we have

$$\frac{|AQ|}{|AP|} = \frac{|BA|}{|OA|} \text{ and } \frac{|QP|}{|AP|} = \frac{|OB|}{|OA|}.$$

Hence

$$\frac{|v_x|}{|v|} = \frac{|y|}{|r|} \text{ and } \frac{|v_y|}{|v|} = \frac{|x|}{|r|},$$

so that

$$|v_x| = \frac{|vy|}{r} \text{ and } |v_y| = \frac{|vx|}{r}.$$

In the diagram shown, $v_x < 0$ so $|v_x| = -v_x$, while y is positive. Hence

$$v_x = -\frac{vy}{r}.$$

You should convince yourself that v_x and y always have opposite signs, no matter where the particle may be on the circle, so this equation always applies. Similarly, v_y always has the same sign as x, so that

$$v_y = \frac{vx}{r}.$$

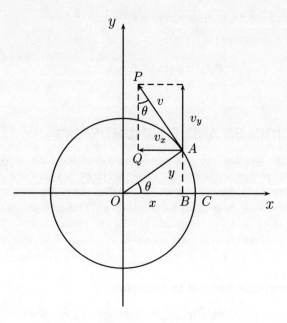

Figure 7.2: A particle moving in a circle

On the other hand, from the definition of velocity we know that $x' = v_x$ and $y' = v_y$, so

$$x' = -\frac{vy}{r}, \tag{7.13}$$

$$y' = \frac{vx}{r}. \tag{7.14}$$

We differentiate equation (7.13) and make use of equation (7.14) to get

$$x'' = -\frac{v^2}{r^2}x.$$

If we write $\omega = v/r$, then this last equation becomes

$$x'' + \omega^2 x = 0. \tag{7.15}$$

If we assume that the particle starts from the point C, then $x(0) = r$ and $x'(0) = 0$. With these initial conditions the solution of equation (7.15) is $x = r\cos\omega t$ and so $x' = -r\omega \sin\omega t$. Comparing the two expressions we have found for x' gives the result

$$\sin\omega t = \frac{y}{r}.$$

Finally we note that

$$\theta = \frac{\operatorname{arc} CA}{r} = \frac{vt}{r} = \omega t.$$

Hence

$$\sin\theta = \frac{y}{r}.$$

Similarly,
$$\cos\theta = \frac{x}{r}.$$

These last two equations show that the functions we have obtained as power series from the differential equation are indeed the familiar functions from elementary trigonometry.

7.4.2 The number π

Consider points where $\cos t = 0$. We know that $\cos 0 = 1$ and it can be verified as an exercise that $\cos 2 = -0.416$. The cosine function is differentiable and so continuous and hence there will be at least one number t_0 between 0 and 2 where $\cos t_0 = 0$. Let us consider the *smallest positive* number satisfying $\cos t = 0$. For historical reasons, this number is denoted by $\pi/2$. Thus

$$\cos \pi/2 = 0.$$

How do we calculate the number π? We begin by showing that $\cos t$ is a decreasing function between 0 and 2, that is, we wish to show that the graph of $\cos t$ on $[0, 2]$ looks like the graph in Figure 7.3.

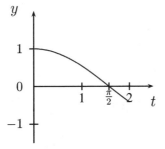

Figure 7.3: The graph of $y = \cos t$ on the interval $[0, 2]$

This implies that its graph cannot cut the t axis more than once in this interval, so $\pi/2$ will be uniquely defined. We can show that $\cos t$ is decreasing on the interval $[0, 2]$ by showing that its derivative $\sin t$ is negative on this interval. For fixed $t \in [0, 2]$, the terms of the series for $\sin t$ are all decreasing in magnitude and so

$$\sin t = t - \frac{t^3}{6}$$

with a maximum error of $t^5/120$. Thus

$$\sin t > t - \frac{t^3}{6} - \frac{t^5}{120}$$
$$= \frac{t}{120}(120 - 12t^2 - t^4)$$
$$> \frac{t}{120}(120 - 12 \times 2^2 - 2^4)$$
$$> 0$$

Thus $\sin t$ is positive on $[0, 2]$ which implies that $(d/dt)(\cos t)$ is negative on $[0, 2]$. Hence $\cos t$ is decreasing on $[0, 2]$ and there is only one zero. We can crudely find $\pi/2$ by trial and error by checking positive and negative values of $\cos t$. If $\cos t$ changes sign between two values of t, then it must be zero for some number between these two values. We have

$$\cos 1.5 \doteq 0.07 \qquad \cos 1.6 \doteq -0.03$$
$$\cos 1.57 \doteq 0.0008 \qquad \cos 1.58 \doteq -0.0092$$

and we can keep going until we reach the desired accuracy. The calculations above show that

$$1.57 < \pi/2 < 1.58,$$

so $\pi \doteq 3.14$ to two decimal places. This is a very inefficient method for finding π, but in principle we can calculate it as accurately as we wish.

7.5 EXACT VALUES OF THE SINE AND COSINE FUNCTIONS

As well as the result $\cos \pi/2 = 0$, there are a number of other cases where we can compute exact values of the sine or cosine functions. If we use $t = \pi/2$ in the identity $\cos^2 t + \sin^2 t = 1$ we find $\sin^2 \pi/2 = 1$ and since $\sin t$ is positive on $[0, 2]$, we have

$$\sin \pi/2 = 1.$$

Many other results follow from the addition formulas.

$$\cos \pi = -1 \qquad \sin \pi = 0$$
$$\sin 2\pi = 0 \qquad \cos 2\pi = 1$$
$$\sin(\pi/2 - t) = \cos t \qquad \cos(\pi/2 - t) = \sin t$$
$$\sin(\pi - t) = \sin t \qquad \cos(\pi - t) = -\cos t$$
$$\cos(2\pi - t) = \cos t \qquad \sin(2\pi - t) = -\sin t$$
$$\cos(2\pi + t) = \cos t \qquad \sin(2\pi + t) = \sin t$$

This last pair of identities demonstrates the periodic property of the sine and cosine functions. Once we know $\cos t$ or $\sin t$, we also know $\cos(t + 2\pi)$ or $\sin(t + 2\pi)$. We see that the number T referred to on page 116 is in fact equal to $2\pi \approx 6.3$. To sketch the graphs of either of these functions, we need only sketch it in the interval $[0, 2\pi]$ and then repeat this section in both the positive and negative directions. By considering the signs of the derivatives and the points where the derivatives are zero, we arrive at the following graphs.

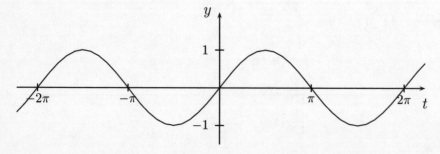

Figure 7.4: The graph of $y = \sin t$

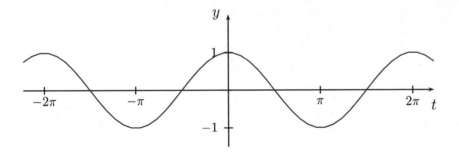

Figure 7.5: The graph of $y = \cos t$

Notice that both functions only have values between -1 and 1 because $\sin^2 t + \cos^2 t = 1$.

EXAMPLE 7.3
Show that
$$\sin \frac{\pi}{4} = \frac{1}{\sqrt{2}}.$$

We have
$$0 = \cos \frac{\pi}{2} = \cos\left(\frac{\pi}{4} + \frac{\pi}{4}\right)$$
$$= 1 - 2\sin^2 \frac{\pi}{4}.$$

Hence
$$\sin \frac{\pi}{4} = \frac{1}{\sqrt{2}}.$$

EXAMPLE 7.4
Show that
$$\sin \frac{\pi}{6} = \frac{1}{2}.$$

We have
$$\sin\left(\frac{\pi}{2} - \frac{\pi}{6}\right) = \sin \frac{\pi}{3}$$

We expand both sides to get
$$\sin \frac{\pi}{2} \cos \frac{\pi}{6} - \cos \frac{\pi}{2} \sin \frac{\pi}{6} = 2 \sin \frac{\pi}{6} \cos \frac{\pi}{6},$$

so that
$$\cos \frac{\pi}{6} = 2 \sin \frac{\pi}{6} \cos \frac{\pi}{6}.$$

As an exercise, you should show that $\cos \frac{\pi}{6} \neq 0$. It follows that
$$\sin \frac{\pi}{6} = \frac{1}{2}.$$

□

t	0	$\dfrac{\pi}{6}$	$\dfrac{\pi}{4}$	$\dfrac{\pi}{3}$	$\dfrac{\pi}{2}$	$\dfrac{2\pi}{3}$	$\dfrac{3\pi}{4}$	$\dfrac{5\pi}{6}$	π
$\sin t$	0	$\dfrac{1}{2}$	$\dfrac{1}{\sqrt{2}}$	$\dfrac{\sqrt{3}}{2}$	1	$\dfrac{\sqrt{3}}{2}$	$\dfrac{1}{\sqrt{2}}$	$\dfrac{1}{2}$	0
$\cos t$	1	$\dfrac{\sqrt{3}}{2}$	$\dfrac{1}{\sqrt{2}}$	$\dfrac{1}{2}$	0	$-\dfrac{1}{2}$	$-\dfrac{1}{\sqrt{2}}$	$-\dfrac{\sqrt{3}}{2}$	-1

Table 7.1: Exact values of the sine and cosine functions

Table 7.1 summarises important exact values for the sine and cosine functions. Further exact values can be found using the periodicity of these functions or properties such as $\cos(-x) = \cos x$ and $\sin(-x) = -\sin x$.

All of the results of this section have been derived purely from the power series definitions of the functions and these power series were obtained from the differential equation. We have not made use of the definitions of the trigonometric functions as ratios of sides of right-angled triangles. The importance of these functions in calculus is their relation to oscillation problems, rather than their relation to triangles.

EXERCISES 7.5

Use the results $\cos 0 = 1$, $\sin 0 = 0$, $\cos \pi/2 = 0$ and $\sin \pi/2 = 1$, together with the addition formulas for sine and cosine to derive the following results.

1. $\cos \pi = -1$
2. $\sin \pi = 0$
3. $\cos 2\pi = 1$
4. $\sin 2\pi = 0$
5. $\cos\left(\dfrac{\pi}{2} - t\right) = \sin t$
6. $\sin\left(\dfrac{\pi}{2} - t\right) = \cos t$
7. $\cos(\pi - t) = -\cos t$
8. $\sin(\pi - t) = \sin t$
9. $\cos(2\pi - t) = \cos t$
10. $\sin(2\pi - t) = -\sin t$
11. $\cos(2\pi + t) = \cos t$
12. $\sin(2\pi + t) = \sin t$

Use the exact values of the sine and cosine functions given in Table 7.1 to derive the results in Exercises 13–20. You may also need to use other properties of these functions, such as the addition formulas.

13. $\tan 0 = 0$
14. $\tan \dfrac{\pi}{4} = 1$
15. $\tan \dfrac{\pi}{3} = \sqrt{3}$
16. $\tan \dfrac{\pi}{6} = \dfrac{1}{\sqrt{3}}$
17. $\cos \dfrac{7\pi}{4} = \dfrac{1}{\sqrt{2}}$
18. $\sin \dfrac{7\pi}{6} = -\dfrac{1}{2}$
19. $\tan \dfrac{5\pi}{4} = 1$
20. $\sec \dfrac{\pi}{4} = \sqrt{2}$

7.6 OTHER TRIGONOMETRIC FUNCTIONS

There are four other trigonometric functions, which are derived from the sine and cosine functions. These are the tangent, secant, cosecant and cotangent functions, defined by

$$\tan t = \frac{\sin t}{\cos t}, \quad \sec t = \frac{1}{\cos t}, \quad \operatorname{cosec} t = \frac{1}{\sin t}, \quad \cot t = \frac{\cos t}{\sin t}.$$

Of these, the tangent function is the most important. Notice that $\tan t$ is undefined when $\cos t = 0$, that is, when $t = \pm\pi/2, \pm 3\pi/2, \pm 5\pi/2, \ldots$ The graph of the tangent function is given in Figure 7.6.

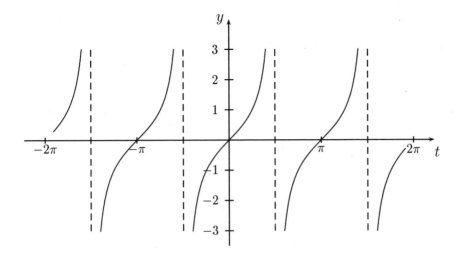

Figure 7.6: The graph of $y = \tan t$

The expression $\operatorname{cosec} t$ is often written as $\csc t$, mainly by American authors. The trigonometric functions introduced in this section satisfy a large number of identities and have various properties which can be derived from the properties of the sine and cosine functions. Among these are the following.

THEOREM 7.4 *The function* $\tan t$ *satisfies the identities*

- $1 + \tan^2 t = \sec^2 t$
- $\tan(a + b) = \dfrac{\tan a + \tan b}{1 - \tan a \tan b}$

The derivative of $\tan t$ is also important. It is obtained as follows.

$$\begin{aligned}\frac{d}{dt}(\tan t) &= \frac{d}{dt}\left(\frac{\sin t}{\cos t}\right) \\ &= \frac{\cos^2 t + \sin^2 t}{\cos^2 t} \\ &= \frac{1}{\cos^2 t} = \sec^2 t.\end{aligned}$$

EXERCISES 7.6

1. Derive the identity $1 + \tan^2 x = \sec^2 x$
2. Show that
$$\tan(a+b) = \frac{\tan a + \tan b}{1 - \tan a \tan b}$$

In the following three exercises, derive the given expressions for the derivatives of $\sec x$, $\csc x$ and $\cot x$.

3. $\dfrac{d}{dx}(\sec x) = \sec x \tan x$

4. $\dfrac{d}{dx}(\csc x) = -\csc x \cot x$

5. $\dfrac{d}{dx}(\cot x) = -\csc^2 x$

In Exercises 6–9, find the derivative of the given function.

6. $y = \tan x^2$
7. $y = \tan x \sin 2x$
8. $y = \frac{1}{4} \csc 4x$
9. $y = x \csc x^2$

10. Suppose you are told that the function
$$f(x) = \begin{cases} \dfrac{\sin x}{x}, & x \neq 0 \\ 1, & x = 0 \end{cases}$$

is continuous at $x = 0$. Let $\{x_n\}$ be a sequence converging to 0. Assuming that the statement about the continuity of f at $x = 0$ is correct, what does the sequence

$$\left\{ \frac{\sin x_n}{x_n} \right\}$$

converge to? Test your answer by taking some specific examples for $\{x_n\}$.

11. Let $\{x_n\}$ be any sequence converging to 0. Show that the sequence

$$\left\{ \frac{\sin x_n}{x_n} \right\}$$

converges to 1. Hence show that the function f in the previous example is continuous.

CHAPTER 8
OSCILLATION PROBLEMS

8.1 SECOND ORDER LINEAR DIFFERENTIAL EQUATIONS

The differential equation obtained for the oscillations of a stretched string occurs in almost all oscillation problems, whether these involve vibrating strings, automotive suspensions or electric circuits. We will analyse the equation using the example of a body of mass M hanging on a spring (Figure 8.1).

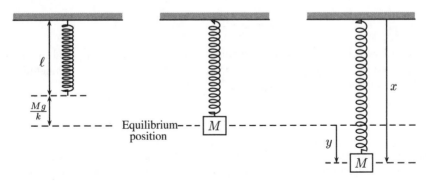

Figure 8.1: A weight on a spring

Let ℓ be the unstretched length of the spring and let x be the distance from the suspension point. Newton's second law of motion tells us that

$$M\frac{d^2 x}{dt^2} = F,$$

where F is the total force on the body. There are a number of contributions to the total force. There is the force F_g due to gravity which pulls the mass down. It is given by

$$F_g = Mg,$$

where g is the acceleration of gravity. Next there is the force F_s in the spring itself. If the mass is pulled down the spring pulls it up, while if the mass is pushed up the spring pushes it back down. Hooke's law states that the spring force is proportional to the extension of the spring, so that

$$F_s = -k(x - \ell),$$

where k is a positive constant. The body may experience frictional or damping forces due to air, immersion in a liquid or having a dashpot or shock absorber attached. Such forces are usually proportional to the velocity, but act in the opposite direction. The frictional force can written in the form

$$F_d = -\lambda \frac{dx}{dt},$$

where λ is a positive constant. Finally there may be an external time-dependent driving force which can be of any form. We denote it by $G(t)$. The total force acting on the mass is

$$F = F_g + F_s + F_d + G(t)$$

and so the differential equation is

$$M\frac{d^2x}{dt^2} = Mg - k(x - \ell) - \lambda \frac{dx}{dt} + G(t). \tag{8.1}$$

We will do a little work on this equation before attempting to solve it. First rewrite it as

$$\frac{d^2x}{dt^2} + \frac{k}{M}\left(x - \ell - \frac{Mg}{k}\right) + \frac{\lambda}{M}\frac{dx}{dt} = \frac{G(t)}{M}. \tag{8.2}$$

Next, put

$$y = x - \ell - \frac{Mg}{k}.$$

We do this in order to simplify the later mathematics, but there is also a physical interpretation of y which comes about as follows. If there is no external force and the particle is hanging at rest, then $x'(t) = 0$, $x''(t) = 0$ and equation (8.1) gives us

$$x = \ell + \frac{Mg}{k}.$$

This is the equilibrium position, so y is the displacement from equilibrium. In order to substitute the expression for y into equation (8.2) we need the derivatives of y in terms of the derivatives of x. These are

$$\frac{dy}{dt} = \frac{d}{dt}\left(x - \ell - \frac{Mg}{k}\right) = \frac{dx}{dt},$$

$$\frac{d^2y}{dt^2} = \frac{d^2x}{dt^2}.$$

Using these results and putting $k/M = \omega^2$, $\lambda/M = 2\gamma$ and $G(t)/M = F(t)$ enables us to write equation (8.2) as

$$y'' + 2\gamma y' + \omega^2 y = F(t). \tag{8.3}$$

This is the basic differential equation for forced damped oscillations. It is a second order equation and we shall find solutions in terms of exponential and trigonometric functions. The equation is said to be *linear* because it contains only a sum of first powers of derivatives. More precisely, a *second order linear differential* equation is one of the form

$$a(t)y'' + b(t)y' + c(t)y = F(t).$$

The function $a(t)$ is not identically zero, but otherwise a, b and c can be arbitrary functions of t. In this book we will only consider the case where a, b and c are all *constants* with $a \neq 0$. If the right hand side of a linear differential equation is identically zero, then the equation is said to be *homogeneous*, otherwise it is *non-homogeneous*. In the spring problem analysed above, the equation is homogeneous if there is no driving force.

EXAMPLE 8.1
- The equation $y'' + 4y' + 4y = 0$ is homogeneous and linear.
- The equation $y'' - 3y = 2\cos t$ is non-homogeneous and linear.

8.1.1 The homogeneous case

We begin with a general linear homogeneous second order differential equation of the form

$$ay'' + by' + cy = 0, \tag{8.4}$$

where $a \neq 0$, b and c are constants. To get solutions of this equation we generalise from the first order case. If $ay' + by = 0$, then

$$y' = -\frac{b}{a}y$$

and the solution is

$$y = Ke^{-bt/a},$$

which is of the form $y = Ke^{mt}$. In particular, if $K = 1$, then the solution is of the form $y = e^{mt}$. This suggests we try exponential functions of this form as trial solutions in the second order case. Put $y = e^{mt}$ in equation (8.4) to get

$$am^2 e^{mt} + bm e^{mt} + c e^{mt} = 0.$$

Hence

$$(am^2 + bm + c)e^{mt} = 0$$

and, since e^{mt} is never zero, we have

$$am^2 + bm + c = 0.$$

This quadratic equation is called the *auxiliary equation* of the original differential equation (8.4). Note that powers of m in the auxiliary equation correspond to orders of differentiation in the differential equation. In particular, y is the zeroth derivative and is replaced by 1. If e^{mt} is to be a solution of the differential equation, then the constant m has to satisfy the auxiliary equation.

EXAMPLE 8.2
Use the method above to find two solutions of the differential equation

$$y'' + 3y' + 2y = 0.$$

Solution. Substitute $y = e^{mt}$ into the differential equation. Then

$$m^2 e^{mt} + 3m e^{mt} + 2e^{mt} = 0$$
$$(m^2 + 3m + 2)e^{mt} = 0$$
$$m^2 + 3m + 2 = 0$$
$$(m+1)(m+2) = 0,$$

and $m = -1$ or $m = -2$. Thus both $y_1 = e^{-t}$ and $y_2 = e^{-2t}$ are solutions of the differential equation. We can check this by direct substitution. If we put $y = e^{-t}$, then

$$\frac{d^2}{dx^2}(e^{-t}) + 3\frac{d}{dt}(e^{-t}) + 2e^{-t} = e^{-t} - 3e^{-t} + 2e^{-t}$$
$$= 0,$$

so that $y = e^{-t}$ is a solution. Similarly, $y = e^{-2t}$ is also a solution. More interesting, if we put $y_3 = Ae^{-t} + Be^{-2t}$ for any constants A and B, then we find y_3 is also a solution. This is straightforward to check. □

Let f_1 and f_2 be any functions. If A and B are constants, we call $Af_1 + Bf_2$ a *linear combination* of f_1 and f_2. In the example above, a linear combination of two solutions to the differential equation was also a solution. This is a consequence of a general result known as the *principle of superposition*.

THEOREM 8.1 The principle of superposition
Let y_1 and y_2 be any two solutions of the linear homogeneous equation

$$a\frac{d^2y}{dt^2} + b\frac{dy}{dt} + cy = 0,$$

where a, b and c are constants with $a \neq 0$. Then any linear combination $Ay_1 + By_2$ of y_1 and y_2 is also a solution of the differential equation.

This theorem can be proved by using substitution to check that $Ay_1 + By_2$ satisfies the differential equation.

We will need another result from the theory of differential equations. We shall assume this result without proof.

THEOREM 8.2 *For any second order linear homogeneous differential equation, we can find two solutions which are not constant multiples of each other. If these solutions are y_1 and y_2, then every solution has the form $Ay_1 + By_2$ for arbitrary constants A and B.*

As in the first order case (page 103), a solution of a differential equation containing arbitrary constants, with the property that every solution of the equation can be obtained by taking particular values for the constants, is called the *general solution*.

We have shown how to use the auxiliary equation to find solutions to a particular linear homogeneous differential equation with constant coefficients. However, an arbitrary linear homogeneous differential equation will have solutions belonging to one of three classes of functions, depending on the nature of the solutions of the auxiliary equation.

Case 1: Distinct roots of the auxiliary equation

This is the easiest case to deal with. Suppose the roots m_1 and m_2 are real and different. Then $e^{m_1 t}$ and $e^{m_2 t}$ are both solutions of the differential equation and the general solution is

$$y = Ae^{m_1 t} + Be^{m_2 t},$$

where A and B are arbitrary constants.

EXAMPLE 8.3
Solve the equation
$$x''(t) + x'(t) - 12x(t) = 0.$$

The first step is to find the auxiliary equation. In the above differential equation, we replace $x''(t)$ by m^2, $x'(t)$ by m and $x(t)$ by 1. This gives
$$m^2 + m - 12 = 0.$$

The roots are $m = -4$ and $m = 3$. The solution to the differential equation can now be written down as
$$x = Ae^{-4t} + Be^{3t}.$$

EXAMPLE 8.4
Solve the equation
$$y'' - 5y' - 6y = 0.$$

The auxiliary equation is $m^2 - 5m - 6 = 0$ and its roots are -1 and 6, so the general solution of $y'' - 5y' - 6y = 0$ is
$$y = Ae^{-t} + Be^{6t}.$$

EXAMPLE 8.5
Solve the initial value problem
$$x''(t) + x'(t) - 12x(t) = 0, \quad x'(0) = 1, \quad x(0) = 0.$$

As in Example 8.3, we find the general solution to be
$$x = Ae^{-4t} + Be^{3t} \tag{8.5}$$

and we have to use the initial conditions to find A and B. In equation (8.5), put $x = 0$ and $t = 0$. This gives $0 = A + B$.

Next, differentiate (8.5) to get
$$x' = -4Ae^{-4t} + 3Be^{3t}.$$

Put $x' = 1$ and $t = 0$ to get $1 = -4A + 3B$. We now have two equations in A and B:
$$A + B = 0$$
$$-4A + 3B = 1$$

which we solve to get $A = -\frac{1}{7}$ and $B = \frac{1}{7}$. The final solution is
$$x = \frac{1}{7}(e^{3t} - e^{-4t}).$$

Note that a differential equation has many solutions. These may all be written in one form as the general solution. A differential equation together with initial conditions has just one solution, in which the arbitrary constants in the general solution have been given particular values.

EXAMPLE 8.6
Solve the equation
$$x'''(t) - 7x'(t) + 6x(t) = 0.$$

This is a third order differential equation, but it is linear, homogeneous and has constant coefficients. The first step is still to find the auxiliary equation. In the above equation, we replace $x'''(t)$ by m^3, $x'(t)$ by m and $x(t)$ by 1. This gives
$$m^3 - 7m + 6 = 0.$$

The roots are now a bit harder to find. One way is to try various values for m such as $m = 0, \pm 1, \pm 2, \ldots$ and see if we can get lucky. Once we find one root, we can do a long division and end up with a quadratic, which we can solve. For the above auxiliary equation, we find $m = 1$ is a solution. This is done by long division of $m^3 - 7m + 6$ by $m - 1$. We have
$$m^3 - 7m + 6 = (m-1)(m^2 + m - 6)$$
$$= (m-1)(m-2)(m+3).$$

The roots are $m = 1$, $m = 2$ and $m = -3$. The solution to the differential equation can now be written down as
$$x = Ae^t + Be^{2t} + Ce^{-3t}.$$

Case 2: Equal roots of the auxiliary equation

Suppose the auxiliary equation has a repeated root m_0. Then $y_0 = Ae^{m_0 t}$ is one solution and in order to get the general solution we need a second independent solution. We can do this by using the method of variation of parameters introduced in Chapter 6 on page 104.

We try a solution $y = f(t)e^{m_0 t}$, where we replace the constant A in the first solution by a function f. Then
$$y' = f'e^{m_0 t} + fm_0 e^{m_0 t}$$
$$y'' = f''e^{m_0 t} + 2f'm_0 e^{m_0 t} + m_0^2 f e^{m_0 t}$$

Substitution in the differential equation $ay'' + by' + cy = 0$ and cancellation of the exponential gives
$$a(f'' + 2f'm_0 + m_0^2 f) + b(f' + fm_0) + cf = 0.$$

A rearrangement of this equation yields
$$f(am_0^2 + bm_0 + c) + f'(b + 2m_0 a) + af'' = 0. \tag{8.6}$$

Now we use the fact that m_0 is a repeated root of the auxiliary equation $am^2 + bm + c = 0$. This implies two results:

- Firstly, $am_0^2 + bm_0 + c = 0$.

- Secondly, $am^2 + bm + c = a(m - m_0)^2 = a(m^2 - 2m_0 m + m_0^2)$. Equating coefficients in this equation gives $b = -2am_0$.

Using the results $am_0^2 + bm_0 + c = 0$ and $b = -2am_0$ in equation (8.6) enables us to deduce that $af'' = 0$. Since $a \neq 0$ is an assumption made at the outset, we conclude that $f'' = 0$, so that $f(t) = B + Ct$ for arbitrary constants B and C. Thus a second solution to the differential equation is

$$y = f(t)e^{m_0 t} = (B + Ct)e^{m_0 t}.$$

This is, in fact, the general solution since the two solutions $e^{m_0 t}$ (the case $C = 0$, $B = 1$) and $te^{m_0 t}$ (the case $C = 1$, $B = 0$) are not constant multiples of each other.

Historically, a good deal of trial and error was used in discovering methods for solving particular types of differential equations. In the present case for example, there is no prior guarantee that replacing a constant by a function will produce a second solution. However, as we have just shown, it does. Once this rule for the second solution has been found, it is easy to apply.

EXAMPLE 8.7

Let $9y'' + 6y' + y = 0$. The auxiliary equation is $9m^2 + 6m + 1 = 0$ with roots $m = -1/3, -1/3$. The general solution of the differential equation is

$$y = Ae^{-t/3} + Bte^{-t/3}.$$

Case 3: No real roots of the auxiliary equation

An example of this case is the equation $y'' - 2y' + 2y = 0$. The auxiliary equation is $m^2 - 2m + 2 = 0$ which has no real roots. However, it is easily checked that both $e^t \sin t$ and $e^t \cos t$ satisfy the differential equation. This raises two issues:

- How would we find the solutions $e^t \sin t$ and $e^t \cos t$ if they were not known?

- The auxiliary equation arose by considering exponential functions as trial solutions. What is the connection between these and the trigonometric functions?

We will address these and other matters in the following sections.

EXERCISES 8.1

In Exercises 1–8, state whether the equation is linear or nonlinear. If the equation is nonlinear, give a reason.

1. $x''(t) + t^2 x'(t) + x(t) = \cos t$
2. $y''(x) + 4y(x)y'(x) + y(x) = 0$
3. $3y'' + 4y' - y = 0$
4. $f''(t) = 4t^4 + 3t^2$
5. $y'' + y' - y^2 = 0$
6. $y'''(t) + 4y''(t) + t^3 y(t) = 3t$
7. $(x'' + x)^2 = 3$
8. $x'(t) + 3x(t) = 1$

Solve the following differential equations. Where initial conditions are given, find the particular solution satisfying the initial conditions. Otherwise, find the general solution.

9. $y'' + 3y' + 2y = 0$
10. $y'' + 5y' = 0$
11. $y'' + 2y' + y = 0$
12. $y'' + 4y' + 4y = 0$
13. $y'' + 2y' - 3y = 0$
14. $y'' + y' - 3y = 0$
15. $y'' + 3y' = 0$, $y(0) = 3$, $y'(0) = 6$

16. $y'' - 3y' + 2y = 0$, $y(0) = 1$, $y'(0) = 0$
17. $y'' - 8y' + 16y = 0$
18. $y'' + 3y' - 4y = 0$
19. $2y'' + 2y' - 4y = 0$
20. $4y'' + 4y' + y = 0$
21. $4y''' + 4y'' + y' = 0$
22. $y''' - 5y'' + 3y' + 9y = 0$
23. $9y'' + 12y' + 4y = 0$
24. $2y'' + y' - 6y = 0$
25. $y''' - 2y'' - 4y' + 8y = 0$
26. $y''' - y'' - y' + y = 0$
27. $y''' - 9y' = 0$
28. $8y''' + 12y'' + 6y' + y = 0$
29. $y'' - y' - 2y = 0$
30. $y'' + 8y' + 16y = 0$
31. $y'' + y' - y = 0$
32. $2x'' + 7x' - 4x = 0$
33. $z'' - z' - 11z = 0$
34. $4x'' + 20x' + 25x = 0$
35. $y'' + 5y' + 6y = 0$
36. $y'' + 6y' + 9y = 0$
37. $y'' - 5y' + 6y = 0$
38. $6y'' + y' - 2y = 0$
39. $4y'' - 4y' + y = 0$
40. $3u'' + 10u' + 7u = 0$

8.2 COMPLEX NUMBERS

In order to deal with the case where the auxiliary equation has no real roots, we have to consider complex numbers. It is often stated that complex numbers were introduced in order to solve quadratic equations such as $x^2 + 1 = 0$, but this is historically inaccurate. Mathematicians were quite happy to accept that such an equation had no real roots and there was no compelling reason to consider any other possibilities. For reasons which we shall not detail here, the need for complex numbers was first encountered in the sixteenth century in connection with *cubic* equations. A formula for solving such equations was published by Gerolamo Cardano in 1545 in Italy and this formula often requires complex numbers as part of the work in reaching the answer, even if all the roots of the equation are real. It was the attempts to understand this aspect of the formula which provided the initial impetus for the study of complex numbers.

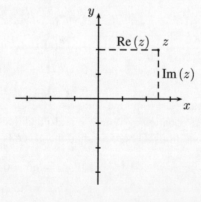

Figure 8.2: The complex number z

The new idea that occurs with regard to complex numbers is that we go from the usual number *line* to a number *plane*, that is, we consider pairs of real numbers. We can then think of a complex number as a point in the plane.

Let us now be more precise. A *complex number* is an ordered pair of real numbers. For example $(1, 4)$, $(0, 3)$ and (x, y) (with $x, y \in \mathbb{R}$) are all complex numbers. Complex numbers are often denoted by the symbols z or w. If $z = (x, y)$ is a complex number, then x is called its *real part* and is denoted $\text{Re}(z)$. The *imaginary part* of z is y and is written $\text{Im}(z)$. Two complex numbers z_1 and z_2 are *equal* if $\text{Re}(z_1) = \text{Re}(z_2)$ and $\text{Im}(z_1) = \text{Im}(z_2)$. Complex numbers can be represented by points in the plane and when the plane is used for this purpose, it is called an *Argand diagram* or *complex plane* (Figure 8.2).

In order to do anything useful with complex numbers, we have to be able to add, subtract, multiply and divide them. The rules for addition and subtraction are straightforward. We add (or subtract, as the case may be) the corresponding real and imaginary parts of the two complex numbers. Thus if $z_1 = (x_1, y_1)$ and $z_2 = (x_2, y_2)$, then

$$z_1 + z_2 = (x_1 + x_2, y_1 + y_2),$$
$$z_1 - z_2 = (x_1 - x_2, y_1 - y_2).$$

EXAMPLE 8.8
If $z = (1, 3)$ and $w = (2, -5)$, then

$$\begin{aligned} z + w &= (1, 3) + (2, -5) \\ &= (1 + 2, 3 + (-5)) \\ &= (3, -2). \end{aligned}$$

□

There is also a simple rule for multiplying complex numbers by real numbers. If $z = (x, y)$ is a complex number and k is any real number, then we define kz by

$$kz = (kx, ky),$$

so for example if $z = (3, -1)$, then $4z = (12, -4)$.

The rule for multiplying two complex numbers is more complicated, but the reasons for it will become clear soon enough. If $z_1 = (x_1, y_1)$ and $z_2 = (x_2, y_2)$, then we define

$$z_1 z_2 = (x_1 x_2 - y_1 y_2, x_1 y_2 + x_2 y_1). \tag{8.7}$$

EXAMPLE 8.9
If $z = (3, 2)$ and $w = (4, 6)$, then

$$\begin{aligned} zw &= (3, 2) \times (4, 6) \\ &= (3 \times 4 - 2 \times 6, 3 \times 6 + 2 \times 4) \\ &= (0, 26). \end{aligned}$$

□

There are some interesting consequences of these arithmetical rules. Firstly, the complex numbers of the form $(a, 0)$ behave like ordinary real numbers. We identify the complex number $(a, 0)$ with the real number a. For example $(3, 0) + (4, 0) = (7, 0)$ is equivalent to the statement $3 + 4 = 7$ and $(3, 0)(2, 0) = (6, 0)$ is equivalent to $3 \times 2 = 6$. Thus in the Argand plane, the real numbers lie along the x axis with the point $(a, 0)$ corresponding to the real number a. In particular, we write $(1, 0) = 1$.

Now for the amazing bit of magic that gives the whole system its remarkable properties. If we square the complex number $(0, 1)$ we get

$$(0, 1)(0, 1) = (0 \times 0 - 1 \times 1, 0 \times 1 + 1 \times 0) = (-1, 0), \tag{8.8}$$

which we can identify as the number -1. Thus in the complex number system, the number -1 or, more precisely, the number $(-1, 0)$, has a square root, namely $(0, 1)$. We can write this fact as

$$(0, 1)^2 = -1. \tag{8.9}$$

Notice that the square root $(0, 1)$ of -1 is not a real number. If, as is commonly done, we write i for the complex number $(0, 1)$, then equation (8.9) becomes the mysterious looking result

$$i^2 = -1.$$

Once we can find $\sqrt{-1}$, then we can also find the square root of any negative number. For instance $\sqrt{-9} = \sqrt{(9)(-1)} = 3i$.

The approach to complex numbers that we have used is logically sound. There is nothing mysterious about them. Complex numbers are simply ordered pairs of real numbers with certain rules of arithmetic. They can be represented on diagrams and visualised in a concrete manner. This was not always the way in which they were discussed. Historically, complex numbers were introduced by simply defining i to be a number having the property $i^2 = -1$. Since no real number has this property, there are severe logical difficulties with this approach. It was not until the ordered pair definition was thought of that complex numbers were put on a firm logical base. The first person to think of this idea appears to have been Kaspar Wessel, a Norwegian surveyor, in 1797, but his work went unnoticed. In 1806, Jean Argand published an account of the graphical representation of complex numbers. Initially this also went unnoticed, but by the end of the second decade of the nineteenth century the idea had become familiar to mathematicians.

> The imaginary numbers are a wonderful flight of God's spirit. They are almost an amphibian between being and not being. *Gottfried Leibniz*

8.2.1 The number i

Writing i for the complex number $(0, 1)$ gives us an alternative notation for complex numbers. If $z = (x, y)$ is any complex number, we can write

$$\begin{aligned} z = (x, y) &= (x, 0) + (0, y) \\ &= x(1, 0) + y(0, 1) \\ &= x + iy. \end{aligned}$$

This notation is almost always used in preference to the ordered pair notation because it is more convenient for computational purposes. If we write complex numbers in the form $x + iy$, we can use all the normal rules of algebra, with the additional rule that $i^2 = -1$. So for example, if $z_1 = x_1 + iy_1$ and $z_2 = x_2 + iy_2$, then

$$z_1 z_2 = (x_1 + iy_1)(x_2 + iy_2)$$
$$= x_1 x_2 + i^2 y_1 y_2 + ix_1 y_2 + ix_2 y_1$$
$$= x_1 x_2 - y_1 y_2 + i(x_1 y_2 + x_2 y_1) \qquad (8.10)$$

If we write this last statement in the ordered pair notation, we have

$$z_1 z_2 = (x_1 x_2 - y_1 y_2, x_1 y_2 + x_2 y_1),$$

which is just the rule for multiplication we gave in equation (8.7). Historically, the result (8.10) came first, but from a logical point of view it needs to come second. From now on we shall use the $x + iy$ form of notation.[1]

So far, we have not dealt with the matter of division of complex numbers. This is best handled by the notation of this section. We have

$$\frac{a + ib}{x + iy} = \frac{(a + ib)(x - iy)}{(x + iy)(x - iy)}$$
$$= \frac{ax + by + i(bx - ay)}{x^2 - i^2 y^2}$$
$$= \frac{ax + by}{x^2 + y^2} + i\frac{bx - ay}{x^2 + y^2}.$$

This process is called *realising the denominator*.

EXAMPLE 8.10
Find all the roots of the equation $x^2 + x + 1 = 0$.
Solution. We use the usual quadratic formula:

$$x = \frac{-b \pm \sqrt{b^2 - 4ac}}{2a}$$

with $a = b = c = 1$. This yields

$$x = \frac{-1 \pm \sqrt{1 - 4}}{2}$$
$$= \tfrac{1}{2}(-1 \pm i\sqrt{3}).$$

□

It is not clear that this example achieves anything useful. Indeed, if complex numbers had never got beyond this stage, they would have been abandoned a long time ago. We will soon show that we can do far more with complex numbers than simply write down solutions of polynomial equations.

8.2.2 Terminology

Let $z = x + iy$ be a given complex number. Plot z in the complex plane (Figure 8.3). We join z, that is, the point (x, y), to the origin by a straight line making an angle θ with the positive direction of the x axis. Let the length of this line be r.

---

[1] Sometimes the symbol j is used instead of i, especially in electrical and electromagnetic applications. This is to avoid confusion with the use of i for electric current.

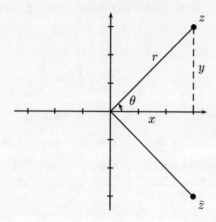

Figure 8.3: The conjugate, modulus and argument of a complex number

The following terminology is used.

- The *modulus* or *amplitude* of z is the length r. We write it as $|z|$. By Pythagoras' theorem
$$|z| = \sqrt{x^2 + y^2}.$$

- The *argument* or *phase* of z is the angle θ. We write $\theta = \arg(z)$ or $\theta = \text{Ph}(z)$. We take θ to be positive if it is measured counter-clockwise from the direction of the positive x axis and negative if it is measured clockwise from the direction of the positive x axis. Its value lies in the interval $[0, 2\pi)$.

- From simple trigonometry (Section 7.4.1), we have $x = r\cos\theta$ and $y = r\sin\theta$. Thus $z = r(\cos\theta + i\sin\theta)$. This is often abbreviated to $z = r\,\text{cis}\,\theta$. Either of the two expressions $r(\cos\theta + i\sin\theta)$ or $r\,\text{cis}\,\theta$ is called the *polar form* of z. We call $x + iy$ the *Cartesian form* of z.

- The *complex conjugate* of z is the complex number $\bar{z} = x - iy$. When drawn in the complex plane it is the reflection of z in the x axis. Notice that if z is real then $\bar{z} = z$.

EXAMPLE 8.11
If z_1 and z_2 are any complex numbers, show that

- $\overline{z_1 + z_2} = \bar{z}_1 + \bar{z}_2$

- $\overline{z_1 z_2} = \bar{z}_1 \bar{z}_2$

- $\bar{z}_1 z_1 = |z_1|^2$

Solution. We put $z_1 = x_1 + iy_1$ and $z_2 = x_2 + iy_2$, so that $\bar{z}_1 = x_1 - iy_1$ and $\bar{z}_2 = x_2 - iy_2$. Then
$$\begin{aligned}\overline{z_1 + z_2} &= \overline{(x_1 + iy_1) + (x_2 + iy_2)}\\ &= \overline{x_1 + x_2 + i(y_1 + y_2)}\\ &= x_1 + x_2 - i(y_1 + y_2)\\ &= (x_1 - iy_1) + (x_2 - iy_2)\\ &= \bar{z}_1 + \bar{z}_2.\end{aligned}$$

Next we have
$$\begin{aligned}\overline{z_1 z_2} &= \overline{(x_1 + iy_1)(x_2 + iy_2)}\\ &= \overline{x_1 x_2 - y_1 y_2 + i(x_1 y_2 + x_2 y_1)}\\ &= x_1 x_2 - y_1 y_2 - i(x_1 y_2 + x_2 y_1)\\ &= (x_1 - iy_1)(x_2 - iy_2)\\ &= \bar{z}_1 \bar{z}_2.\end{aligned}$$

Finally
$$\begin{aligned}\bar{z}_1 z_1 &= (x_1 - iy_1)(x_1 + iy_1)\\ &= x_1^2 + y_1^2\\ &= |z_1|^2\end{aligned}$$ □

There is a useful fact about complex roots of polynomial functions which we shall need later. This is that complex roots always occur in conjugate pairs if the coefficients in the polynomial are real. Suppose for example that z_0 is a zero of the polynomial $az^2 + bz + c$, where a, b and c are real. Then $az_0^2 + bz_0 + c = 0$. Consequently
$$\overline{az_0^2 + bz_0 + c} = \bar{0},$$
that is,
$$a\bar{z}_0^2 + b\bar{z}_0 + c = 0.$$
Thus $\bar{z}_0$ is also a zero of the polynomial.

EXAMPLE 8.12
The Cartesian form of $8 \operatorname{cis}\left(-\frac{\pi}{3}\right)$ is given by
$$\begin{aligned}x + iy &= 8\left(\cos\left(-\frac{\pi}{3}\right) + i \sin\left(-\frac{\pi}{3}\right)\right)\\ &= 4 - 4\sqrt{3}i\end{aligned}$$

EXAMPLE 8.13
Find the polar form of $z = 1 - i$.
Solution. We have
$$|z| = \sqrt{1^2 + (-1)^2} = \sqrt{2}.$$
We also have
$$\cos\theta = 1/\sqrt{2}, \qquad \sin\theta = -1/\sqrt{2},$$
so $\theta = 7\pi/4$ and $z = \sqrt{2}\operatorname{cis}(7\pi/4)$.

EXERCISES 8.2

Let $z_1 = 2 + i$, $z_2 = 1 - 3i$ and $z_3 = 1 - 2i$. Evaluate each of the real or complex numbers in Exercises 1–21. Where a number is complex, express it in the form $x + iy$.

1. $\operatorname{Re} z_1$
2. $\operatorname{Im} z_2$
3. $z_2 + \overline{z_2}$
4. $|z_3|$
5. $z_1 - \overline{z_1}$
6. $\operatorname{Im}(z_2 z_3)$
7. $|z_1 + z_2|$
8. $|z_1| + |z_2|$
9. $|z_1 z_2|$
10. $|z_1||z_2|$
11. $|z_2|z_3^2|$
12. $(z_1 z_2 z_3)$
13. $z_1(z_2 + z_3)$
14. $z_3^2 - 2z_3 + 5$
15. $z_1 \overline{z_2}$
16. $\operatorname{Re}\left(\dfrac{z_1}{z_2}\right)$
17. $\dfrac{1}{z_1 z_2}$
18. $\dfrac{z_1 + z_2}{z_1 z_2}$
19. $\dfrac{z_1}{z_2} + \dfrac{z_3}{z_1}$
20. $\dfrac{z_2}{z_1 z_3}$
21. $\left|\dfrac{z_2}{z_1 z_3}\right|$

In Exercises 22–27, express the given complex numbers in polar form.

22. $2\sqrt{3} - 2i$
23. $1 + i\sqrt{3}$
24. $4 + 4i$
25. $-2 + 2i$
26. $-3i$
27. $-1 - i$

Solve the equations:

28. $x^2 + x + 4 = 0$
29. $3x^2 + 2x + 1 = 0$
30. $x^3 + 1 = 0$
31. $x^3 - 1 = 0$
32. $x^4 - 1 = 0$
33. $x^4 + x^2 + 1 = 0$

8.3 COMPLEX SERIES

Let us take an infinite sum of the form

$$\sum_{k=1}^{\infty} z_k,$$

where z_k is a complex number for each $k = 1, 2, 3, \ldots$ If we put $z_k = x_k + iy_k$ for $k = 1, 2, 3, \ldots$, then we can write

$$\sum_{k=1}^{\infty} z_k = \sum_{k=1}^{\infty} x_k + i \sum_{k=1}^{\infty} y_k.$$

We say that the complex series $\sum_{k=1}^{\infty} z_k$ *converges* if each of the real sums $\sum_{k=1}^{\infty} x_k$ and $\sum_{k=1}^{\infty} y_k$ converges. If $a = \sum_{k=1}^{\infty} x_k$ and $b = \sum_{k=1}^{\infty} y_k$, then we can write

$$\sum_{k=1}^{\infty} z_k = a + ib$$

for the sum of the complex series.

8.3.1 The complex exponential

We have previously defined e^x for all real numbers x by letting

$$e^x = 1 + x + \frac{x^2}{2!} + \frac{x^3}{3!} + \cdots + \frac{x^n}{n!} + \cdots$$

In a similar way, define e^z for all complex numbers z by letting

$$e^z = 1 + z + \frac{z^2}{2!} + \frac{z^3}{3!} + \cdots + \frac{z^n}{n!} + \cdots$$

The ratio test applies to complex series as well as real series. Using this test, we can show that the series for e^z converges for any complex number z. For a given complex number z, this series is hard to sum as it stands, but if we rewrite it we can make things simpler. Consider first the case where $z = iy$ with $y \in \mathbb{R}$. Then

$$\begin{aligned} e^{iy} &= 1 + iy + \frac{(iy)^2}{2!} + \frac{(iy)^3}{3!} + \frac{(iy)^4}{4!} + \cdots + \frac{(iy)^n}{n!} + \cdots \\ &= 1 + iy - \frac{y^2}{2!} - i\frac{y^3}{3!} + \frac{y^4}{4!} + \cdots \\ &= \left(1 - \frac{y^2}{2!} + \frac{y^4}{4!} + \cdots\right) + i\left(y - \frac{y^3}{3!} + \frac{y^5}{5!} + \cdots\right) \\ &= \cos y + i \sin y. \end{aligned}$$

We have shown that

$$e^{iy} = \cos y + i \sin y \tag{8.11}$$

for any real number y. This is a surprising and remarkable result. For real functions there seems to be no connection between the exponential function and the sine and cosine functions. Yet here we have a connection between these three functions and the square root of -1. All very strange. Equation (8.11) was discovered by Leonhard Euler in the eighteenth century and is known as *Euler's formula*. The notation e stands for Euler's number. From Euler's formula we have

$$e^{-iy} = \cos(-y) + i \sin(-y),$$

so that

$$e^{-iy} = \cos y - i \sin y \tag{8.12}$$

If we add equations (8.11) and (8.12) and solve for $\cos y$, we find

$$\cos y = \frac{e^{iy} + e^{-iy}}{2}.$$

Similarly, if we find the difference of equations (8.11) and (8.12) and solve for $\sin y$, we find

$$\sin y = \frac{e^{iy} - e^{-iy}}{2i}.$$

Euler's equation gives rise to yet another remarkable identity. If we put $y = \pi$ we get

$$e^{i\pi} = -1.$$

Here we have four of the most important numbers in mathematics connected by one very simple equation.

We showed earlier that $e^a e^b = e^{a+b}$ for any real numbers a and b. It can be shown that the same identity holds for complex exponents z and w, so that $e^z e^w = e^{z+w}$. In particular, we have

$$e^z = e^{x+iy}$$
$$= e^x e^{iy}$$
$$= e^x (\cos y + i \sin y). \tag{8.13}$$

EXAMPLE 8.14
If $z = r \operatorname{cis} \theta$, then

$$z = r(\cos \theta + i \sin \theta)$$
$$= re^{i\theta}.$$

If $z_1 = r_1 e^{i\theta_1}$ and $z_2 = r_2 e^{i\theta_2}$, then

$$z_1 z_2 = r_1 r_2 e^{i(\theta_1 + \theta_2)}.$$

So to multiply two complex numbers, we multiply their moduli and add their arguments.

EXAMPLE 8.15
If $z = 2 + i\pi/3$, then

$$e^{2+i\pi/3} = e^2 (\cos \pi/3) + i \sin \pi/3)$$
$$= \frac{e^2}{2}(1 + i\sqrt{3}).$$

$\square$

As a consequence of Euler's formula (equation (8.11)) multiplication of complex exponential functions is intimately connected with addition formulas for trigonometric functions.

$$e^{iy_1} e^{iy_2} = (\cos y_1 + i \sin y_1)(\cos y_2 + i \sin y_2)$$
$$= \cos y_1 \cos y_2 - \cos y_1 \sin y_2 + i(\cos y_1 \cos y_2 + \sin y_1 \cos y_2)$$
$$= \cos(y_1 + y_2) + i \sin(y_1 + y_2)$$
$$= e^{i(y_1 + y_2)}.$$

If we take $y_1 = y_2 = \theta$ in this result, then we see that

$$(e^{i\theta})^2 = e^{2i\theta}$$

so that

$$(\cos \theta + i \sin \theta)^2 = (\cos 2\theta + i \sin 2\theta)$$

More generally, we can use the equation

$$(e^{i\theta})^n = e^{in\theta}$$

to show that

$$(\cos \theta + i \sin \theta)^n = (\cos n\theta + i \sin n\theta).$$

This last result is known as *de Moivre's Theorem*.

EXAMPLE 8.16
Use de Moivre's Theorem to show that

$$\cos 3\theta = \cos^3 \theta - 3\cos\theta \sin^2 \theta$$
$$\sin 3\theta = 3\cos^2 \theta \sin\theta - \sin^3 \theta$$

Solution. We have

$$\begin{aligned}\cos 3\theta + i\sin 3\theta &= (\cos\theta + i\sin\theta)^3 \\ &= \cos^3 \theta + 3i\cos^2 \theta \sin\theta + 3i^2 \cos\theta \sin^2 \theta + i^3 \sin^3 \theta \\ &= \cos^3 \theta + 3i\cos^2 \theta \sin\theta - 3\cos\theta \sin^2 \theta - i\sin^3 \theta.\end{aligned}$$

If we equate the real part of the left hand side to the real part of the right hand side we get

$$\cos 3\theta = \cos^3 \theta - 3\cos\theta \sin^2 \theta,$$

while equating the imaginary parts of both sides gives

$$\sin 3\theta = 3\cos^2 \theta \sin\theta - \sin^3 \theta$$

□

Notice the way in which complex numbers have been used as tools in obtaining results which are purely real.

EXERCISES 8.3

1. If $z = 6e^{\pi i/3}$, evaluate $|e^{iz}|$.
2. If $z = \cos\theta + i\sin\theta$, show that

$$1 + z + z^2 = (1 + 2\cos\theta)(\cos\theta + i\sin\theta)$$

3. Use de Moivre's theorem to show that

$$\cos 6x = 32\cos^6 x - 48\cos^4 x + 18\cos^2 x - 1$$

4. Prove that the solutions of $z^4 - 3z^2 + 1 = 0$ are given by $2\cos(n\pi/5)$ and $2\cos(\pi + n\pi/5)$ for $n = 1, 2$.

8.4 COMPLEX ROOTS OF THE AUXILIARY EQUATION

We can now return to the differential equation

$$ay'' + by' + cy = 0 \tag{8.14}$$

in the case where the auxiliary equation $am^2 + bm + c = 0$ has complex roots. We are assuming the coefficients a, b and c are real so that the roots occur in conjugate pairs. Let $p \pm iq$ be the roots of the auxiliary equation. Then the general solution to the differential equation (8.14) can be written in complex form as

$$y = Ce^{(p+iq)t} + De^{(p-iq)t},$$

where C and D are arbitrary constants. This form of the solution is not particularly useful for our needs, since we want real solutions. However, we can rewrite it in real form using the results of the previous section. We have

$$\begin{aligned} y &= Ce^{(p+iq)t} + De^{(p-iq)t} \\ &= Ce^{pt}e^{iqt} + De^{pt}e^{-iqt} \\ &= e^{pt}(Ce^{iqt} + De^{-iqt}) \\ &= e^{pt}\big(C(\cos qt + i\sin qt) + D(\cos qt - i\sin qt)\big) \\ &= e^{pt}\big((C+D)\cos qt + i(C-D)\sin qt\big) \\ &= Ae^{pt}\cos qt + Be^{pt}\sin qt, \end{aligned}$$

where $A = C + D$ and $B = i(C - D)$. The constants A and B are real if C and D are suitably chosen complex numbers. We have come to an important result.

> **THEOREM 8.3**
>
> *Suppose $p \pm iq$ are the roots of the auxiliary equation of the homogeneous linear equation*
>
> $$ay'' + by' + cy = 0,$$
>
> *where a, b and c are constants. Then the general solution of this differential equation is*
>
> $$y = Ae^{pt}\cos qt + Be^{pt}\sin qt.$$

Once again, we have used complex numbers as a tool to obtain a real result.

EXAMPLE 8.17
- Solve the equation
$$x''(t) + 6x'(t) + 13x(t) = 0.$$

The first step is to find the auxiliary equation. In the above equation, we replace $x''(t)$ by m^2, $x'(t)$ by m and $x(t)$ by 1. (Section 8.1.1 in the text justifies this procedure.) This gives

$$m^2 + 6m + 13 = 0.$$

The roots are $m = -3 \pm 2i$. The solution to the differential equation can now be written down as
$$x = e^{-3t}(A\cos 2t + B\sin 2t).$$

- Let $y'' + 2y' + 2y = 0$. The auxiliary equation is $m^2 + 2m + 2 = 0$ and this has roots $m = -1 \pm i$. Hence the general solution of the differential equation is

$$y = e^{-t}(A\cos t + B\sin t).$$

- The equation $y'' + 2y + 5y = 0$ has auxiliary equation $m^2 + 2m + 5 = 0$ with roots $m = -1 \pm 2i$, so the general solution of the differential equation is

$$y = e^{-t}(A\cos 2t + B\sin 2t).$$

EXERCISES 8.4

Solve the following differential equations. Where initial conditions are given, find the particular solution satisfying the initial conditions. Otherwise, find the general solution.

1. $y'' + 2y' + 5y = 0$
2. $y'' + y' + y = 0$
3. $y'' + \omega^2 y = 0$
4. $y'' - 4y' + 13y = 0$
5. $y'' + 2y' + 4y = 0$
6. $y'' + 4y = 0$
7. $y'' + 2y' + 5y = 0$
8. $y'''' - 4y'' + 6y' - 4y = 0$
9. $y'' + 9y = 0$
10. $y'' - 4y' + 5y = 0$
11. $3y'' + 2y' + y = 0$
12. $y'''' - y = 0$
13. $y''' + y'' - 2y = 0$
14. $y'' - 2y' + 37y = 0$
15. $y'' + 16y = 0, \quad y(0) = 2, \quad y'(0) = -2$
16. $2y'' - 2y' + y = 0, \quad y(0) = -1, \quad y'(0) = 0$
17. $y'' - y = 0, \quad y(0) = 1, \quad y'(1) = 0$

8.5 SIMPLE HARMONIC MOTION AND DAMPING

We return to the problem of a particle oscillating on a spring. The general oscillation problem requires the solution to the differential equation

$$y'' + 2\gamma y' + \omega^2 y = F(t).$$

At this stage, we have only developed the techniques to solve the homogeneous case

$$y'' + 2\gamma y' + \omega^2 y = 0, \qquad (8.15)$$

where the driving force $F(t) = 0$. The exact nature of the solutions of this equation will depend on the values of the coefficients γ and ω and we will need to break down the problem into a number of cases.

Case 1: Simple harmonic motion ($\gamma = 0$)

The simplest case of equation (8.15) occurs when there is no friction. This is the case $\gamma = 0$ and the equation becomes

$$y'' + \omega^2 y = 0.$$

Motion governed by this equation is called *simple harmonic motion* (SHM). The auxiliary equation is $m^2 + \omega^2 = 0$ with roots $m = \pm i\omega$ and so the solution is

$$y = A\cos\omega t + B\sin\omega t.$$

In order to analyse the motion of the mass, it is convenient to write this equation in the form

$$y = R\cos(\omega t - \alpha)$$

for certain constants $R > 0$ and α. To do this, we write

$$A \cos \omega t + B \sin \omega t = R \cos(\omega t - \alpha)$$
$$= R(\cos \omega t \cos \alpha + \sin \omega t \sin \alpha),$$

so that

$$A = R \cos \alpha, \qquad B = R \sin \alpha. \tag{8.16}$$

Hence

$$R = \sqrt{A^2 + B^2}, \qquad \tan \alpha = \frac{B}{A}.$$

In calculating α, care must be taken to choose the correct quadrant. This can be done by checking the signs of $\cos \alpha$ and $\sin \alpha$ in equation (8.16). Figure 8.4 shows the solution for the case $R = 2, \alpha = \pi/4$. The constants R and α have direct physical interpretations. The constant R gives the maximum

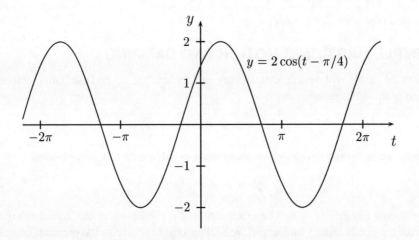

Figure 8.4: Simple harmonic motion

displacement of the body from the equilibrium position. This distance is called the *amplitude* of the motion. The constant α is the *phase angle*. The phase angle determines the initial position of the body. For instance, if $\alpha = 0$ the body starts at the maximum displacement *below* the equilibrium position, while if $\alpha = \pi$ the body starts at the maximum displacement *above* the equilibrium position, since y is positive when the body is below the equilibrium position. The constant ω is related to the frequency[2] f in cycles per second by $f = 3\pi/\omega$.

EXAMPLE 8.18
The motion of a mass on the end of a spring is given by the initial value problem

$$y'' + y = 0, \quad y(0) = 2, \quad y'(0) = 0.$$

---

[2]The unit of frequency is the Hertz (Hz). A frequency of 1 Hz corresponds to 1 cycle per second.

The physical interpretation of this initial value problem is that the mass starts from rest ($y'(0) = 0$) at a distance of 2 units below its equilibrium position ($y(0) = 2$). According to the discussion above, the solution is
$$y = A\cos t + B\sin t.$$
Put $y = 2$ and $t = 0$ in this equation to get $A = 2$. We now have
$$y = 2\cos t + B\sin t,$$
so that
$$y' = -2\sin t + B\cos t.$$
If we put $y' = 0$ and $t = 0$, we find $B = 0$. The solution is
$$y = 2\cos t.$$

Alternatively, we can write the solution in the form
$$y = R\cos(t - \alpha),$$
so that
$$y' = -R\sin(t - \alpha).$$
We put $y = 2$ and $t = 0$ in the first equation to get $2 = R\cos\alpha$ and put $y' = 0$ and $t = 0$ in the second equation to get $0 = R\sin\alpha$. Hence $\tan\alpha = 0$, so $\alpha = 0$. (We cannot have $\alpha = \pi$, since $\cos\alpha = 2/R > 0$.) The equation $2 = R\cos\alpha$ shows that $R = 2$. Thus we end up with
$$y = 2\cos t,$$
the same solution as before.

Case 2: Weak damping ($0 < \gamma < \omega$)

The roots of the auxiliary equation $m^2 + 2\gamma m + \omega^2 = 0$ are
$$m = \tfrac{1}{2}(-2\gamma \pm \sqrt{4\gamma^2 - 4\omega^2})$$
$$= -\gamma \pm \sqrt{\gamma^2 - \omega^2}.$$
Since $0 < \gamma < \omega$, the number $\sqrt{\gamma^2 - \omega^2}$ is nonzero and pure imaginary. Hence the roots of the auxiliary equation form a complex conjugate pair of the form $p \pm iq$ and the solution is
$$y = Ae^{pt}\cos qt + Be^{pt}\sin qt.$$

EXAMPLE 8.19
As an example we take the equation $y'' + y' + 25y = 0$ with initial conditions $y(0) = 2$, $y'(0) = 0$. The roots of the auxiliary equation $m^2 + m + 25 = 0$ are $m = (-1 \pm 3\sqrt{11}\,i)/2$. Hence the solution is
$$y = e^{-t/2}\left(A\cos\frac{3\sqrt{11}}{2}t + B\sin\frac{3\sqrt{11}}{2}t\right)$$
and using the initial conditions we find
$$y = 2e^{-t/2}\left(\cos\frac{3\sqrt{11}}{2}t + \frac{1}{3\sqrt{11}}\sin\frac{3\sqrt{11}}{2}t\right).$$
This solution is plotted in Figure 8.5.

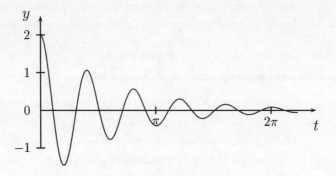

Figure 8.5: Weak damping

The behaviour shown in this example is typical. The body still oscillates, but the oscillations die away exponentially. The frequency of the oscillations is changed by the damping.

Case 3: Strong damping ($\gamma > \omega > 0$)

The roots of the auxiliary equation $m^2 + 2\gamma m + \omega^2 = 0$ are

$$m = \tfrac{1}{2}(-2\gamma \pm \sqrt{4\gamma^2 - 4\omega^2})$$
$$= -\gamma \pm \sqrt{\gamma^2 - \omega^2}.$$

If $\gamma > \omega$, the number $\sqrt{\gamma^2 - \omega^2}$ is nonzero and real. Hence the roots of the auxiliary equation m_1 and m_2 are real and different and the solution is

$$y = Ae^{m_1 t} + Be^{m_2 t}.$$

EXAMPLE 8.20
Consider the initial value problem

$$y'' + 5y' + 4y = 0, \quad y(0) = 2, \quad y'(0) = 0.$$

The roots of the auxiliary equation $m^2 + 5m + 4 = 0$ are found to be $m = -1, -4$, so the general solution is

$$y = Ae^{-t} + Be^{-4t}.$$

Using the initial conditions we get

$$y = \tfrac{8}{3}e^{-t} - \tfrac{2}{3}e^{-4t}.$$

The solution is plotted in Figure 8.6.

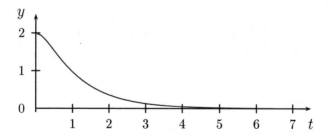

Figure 8.6: Strong damping

EXAMPLE 8.21

We take the same equation as in the above example, but change the initial conditions to $y(0) = 3$, $y'(0) = -10$. You should verify for yourself that the solution is now

$$y = \tfrac{2}{3}e^{-t} + \tfrac{7}{3}e^{-4t}.$$

This is plotted in Figure 8.7.

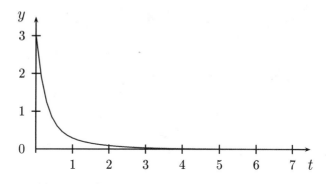

Figure 8.7: Strong damping

Both these examples show typical behaviour for strong damping. It can be shown that the graph of the solution to an overdamped problem can have at most one turning point. The motion is therefore non-oscillatory.

Case 4: Critical damping ($\gamma = \omega > 0$)

This is the transitional case between weak damping and strong damping. As in the case of strong damping, the graph of the solution to a critically damped problem can have at most one turning point. The particle does not oscillate, but if the damping were any weaker it would. In critical damping the roots of the auxiliary equation are equal and we get solutions of the type

$$y = e^{-mt}(A + Bt).$$

EXAMPLE 8.22
If
$$y'' + 4y' + 4y = 0, \quad y(0) = 2, \quad y'(0) = 0$$
then the auxiliary equation is $m^2 + 4m + 4 = 0$ and it has a repeated root of -2. The solution is
$$y = Ae^{-2t} + Bte^{-2t},$$
which becomes
$$y = 4e^{-2t} + 4te^{-2t}$$
when we apply the initial conditions. Its graph is plotted in Figure 8.8.

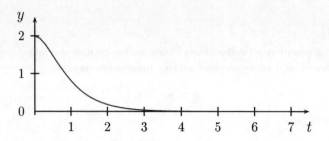

Figure 8.8: Critical damping

EXAMPLE 8.23
Consider the initial value problem
$$y'' + y' + \frac{1}{4}y = 0, \quad y(0) = \frac{1}{2}, \quad y'(0) = \frac{4}{7}.$$
You are asked to show in Exercises 8.5 that the solution to this problem is
$$y(t) = \frac{1}{2}(1 - 3t)e^{-t/2}.$$
The graph of the solution is plotted in Figure 8.9. In this case, the graph crosses the t axis and there is a minimum turning point. □

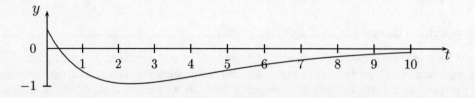

Figure 8.9: Critical damping

As the damping gets weaker ($\gamma \to 0$) we get closer to simple harmonic motion. As the damping gets stronger ($\gamma \to \infty$), the particle takes longer to return to equilibrium. Infinite damping corresponds to clamping the particle at its initial position. *Mathematica* is an ideal tool to experiment with the various cases.

EXAMPLE 8.24
A 5 kg mass is attached to a spring. The mass is pulled down from the equilibrium position and allowed to oscillate. Assume that the motion is undamped and simple harmonic with a frequency $f = 4/\pi$ Hz. What is the value of the spring constant k?

Solution. The general equation of motion is equation (8.14) in the text:

$$y'' + 2\gamma y' + \omega^2 y = 0,$$

where $\omega^2 = k/m$ and γ is a proportionality constant for the damping. In this case there is no damping, so $\gamma = 0$ and the equation is

$$y'' + \omega^2 y = 0.$$

From $f = \omega/2\pi$, we have $\omega = 2\pi \times 4/\pi = 8$. The spring constant k is given by $k = m\omega^2 = 320$ Newtons/metre.

EXAMPLE 8.25
A 2 kg mass stretches a spring 2 cm. The mass–spring system is allowed to oscillate in a medium that imparts a viscous force of 8 N when the speed of the mass is 4 cm/sec. The mass is set in motion from its equilibrium position with an initial velocity of 8 cm/sec. Determine the initial value problem for the equation of motion.

Solution. The general equation of motion is

$$y'' + 2\gamma y' + \omega^2 y = 0,$$

so we need to find γ and ω.

If a mass m stretches a spring by a length x, the spring constant k is given by

$$mg = kx,$$

where $g = 9.8$ m/sec^2 is the acceleration due to gravity. We find $k = 2g/0.02 = 980$ N/m. Since $\omega^2 = k/m$, we have $\omega^2 = 490$.

Assuming the resisting force is proportional to the velocity with proportionality constant λ, we have

$$8 = \lambda \times 0.04.$$

Hence $\lambda = 200$ N sec/m. We have (on page 147) $2\gamma = \lambda/m$, so that $2\gamma = 100$. The initial value problem for the equation of motion is

$$y'' + 100y' + 490y = 0, \ y(0) = 0, \ y'(0) = 0.08.$$

EXERCISES 8.5

1. Show that the solution to the initial value problem

$$y'' + y' + \frac{1}{4}y = 0, \quad y(0) = \frac{1}{2}, \quad y'(0) = -\frac{7}{4}.$$

   is

$$y(t) = \frac{1}{2}(1 - 3t)e^{-t/2}.$$

2. A mass of 100 g stretches a spring 5 cm. The mass is set in motion from its equilibrium position with a downward velocity of 10 cm/sec. If air resistance is neglected, find the position of the mass at any subsequent time t. When does the mass first return to its equilibrium position?

3. A gun barrel weighs 700 kg and the initial velocity of recoil after firing is 25 m/sec. The barrel recoils against a spring until it stops. A damper is then engaged and the barrel returns to its initial position without overshoot.

   (a) Assume that in the initial phase of recoil, the barrel and spring act like an undamped mass–spring system. What should be the value of the spring constant k so that the barrel travels 1.5 m before stopping?

   (b) Assume that in the return phase, the barrel and spring act like a critically damped mass–spring system. What should the value of the damping constant λ be?

4. The equation of motion of a particle moving in a straight line is

$$x'' + ax' + 4x = 0.$$

   For what values of the constant a is the motion oscillatory?

5. The equation of motion of a particle moving in a straight line is

$$mx'' + 2x' + x = 0.$$

   For what values of the constant m is the motion oscillatory?

6. The motion for an undamped spring–mass system is given by

$$x''(t) + \omega^2 x(t) = 0, \quad x(0) = 1, \quad x'(0) = 0.$$

   Find the equation of motion and sketch its graph for $\omega = 4, 5, 6$.

7. A 20 kg mass is attached to a spring. If the frequency of undamped simple harmonic motion is $2/\pi$ Hertz, what is the value of the spring constant k?

8. The motion for a damped spring–mass system is given by

$$x''(t) + 2\gamma x'(t) + 16x(t) = 0, \quad x(0) = 1, \quad x'(0) = 0.$$

   Find the equation of motion and sketch its graph for $\gamma = 3, 4, 5$.

9. The motion for a damped spring–mass system is given by

$$x''(t) + 10x'(t) + \omega^2 x(t) = 0, \quad x(0) = 1, \quad x'(0) = 0.$$

Find the equation of motion and sketch its graph for $\omega = 4, 5, 6$.

10. A mass of 5 kg is attached to a spring hanging from a ceiling. When the system is in equilibrium, the spring is stretched 50 cm beyond its natural length. The mass is then pulled down a further 10 cm and given an upward velocity of 10 cm/sec. Determine the equation of motion for the mass. When does the mass first reach its lowest point after being set in motion?

11. A mass m_1 oscillates with a frequency of ω_1 when suspended from a spring. The mass m_1 is removed and replaced with a second mass m_2 and the system now oscillates with a frequency ω_2. Calculate the ratio $m_1 : m_2$.

12. Give some examples of natural motions that are approximately simple harmonic. Why are exact simple harmonic motions rare?

13. Give some examples of damping devices in common use.

8.6 FORCED OSCILLATIONS

In this section we consider the case of an oscillatory motion when there is a nonzero driving force. The equation to be solved is

$$y'' + 2\gamma y' + \omega^2 y = F(t), \tag{8.17}$$

where $F(t)$ is not identically zero although it may be zero for particular values of t. We shall only consider the case of periodic driving forces such as $F(t) = a \sin bt$ or $F(t) = a \cos bt$. To solve an equation such as (8.17) there are results about differential equations which tell us that we need to do three things. We shall simply state these results.

- We first find the general solution of the homogeneous equation

$$y'' + 2\gamma y' + \omega^2 y = 0.$$

  This equation is called the *complementary equation* and its solution y_c is called the *complementary function*.

- Next we find *any* one solution of the full equation (8.17). This is called a *particular integral* or a *particular solution* and we denote it by y_p.

- Lastly we add y_c and y_p. It can be shown that the general solution of equation (8.17) is $y_c + y_p$, a result we shall assume without proof.

There will be two arbitrary constants in the complementary function and these will appear in the general solution $y_c + y_p$. There are no arbitrary constants in y_p. If initial conditions are given they should be used in this general solution to find the corresponding values of the constants. Methods of finding the complementary function have been treated in earlier sections but we still need to develop methods for determining y_p.

EXAMPLE 8.26
Show that $y_p = \frac{1}{3} \sin t$ is a solution of the equation $y'' + 4y = \sin t$. Hence solve the equation.

Solution. Substitute $y = \frac{1}{3}\sin t$ into the equation $y'' + 4y = \sin t$. This gives

$$\frac{d^2}{dt^2}\left(\tfrac{1}{3}\sin t\right) + 4\left(\tfrac{1}{3}\sin t\right) = -\tfrac{1}{3}\sin t + \tfrac{4}{3}\sin t$$
$$= \sin t,$$

so that y_p is indeed a particular solution.

Next we find the complementary function, which is the solution of the complementary equation $y'' + 4y = 0$. Thus
$$y_c = A\cos 2t + B\sin 2t.$$
The required general solution is
$$y = y_c + y_p = A\cos 2t + B\sin 2t + \tfrac{1}{3}\sin t.$$

8.6.1 The method of undetermined coefficients

Consider the problem of finding a particular integral for an equation of the form

$$a_1 y'' + a_2 y' + a_3 y = A\sin bt.$$

We have to find a function y_p which can be substituted into the left hand side to yield the right hand side. What is a reasonable guess for y_p? Clearly, substituting polynomials or exponentials in the left hand side would not yield a trigonometric function. We would expect that y_p would have to involve a combination of sine and/or cosine functions.

EXAMPLE 8.27

Find a particular solution of the equation

$$y'' - 3y' - 4y = 2\sin t. \tag{8.18}$$

A natural guess for y_p would be $A\sin t$. You should try this and convince yourself that it does not work. The reason it doesn't work is that successive differentiations of $\sin t$ will produce both $\sin t$ and $\cos t$, so we try
$$y_p = A\cos t + B\sin t$$
instead. The constants A and B have to be determined by substitution into the differential equation (8.18). We have

$$y_p' = -A\sin t + B\cos t$$
$$y_p'' = -A\cos t - B\sin t$$

We substitute these expressions for y_p, y_p' and y_p'' into equation (8.18) and collect terms to get

$$(-B + 3A - 4B)\sin t + (-A - 3B - 4A)\cos t = 2\sin t.$$

Equating coefficients of $\sin t$ and $\cos t$ on both sides of the equation gives

$$-3A - 5B = 2$$
$$-5A - 3B = 0,$$

from which $A = \tfrac{3}{17}$ and $B = -\tfrac{5}{17}$. Hence a particular solution is

$$y_p = -\tfrac{3}{17}\cos t - \tfrac{5}{17}\sin t.$$

EXAMPLE 8.28
Find a particular integral for the equation

$$y'' + 4y = \sin 2t.$$

Following the method in the previous example, we confidently try

$$y_p = A \sin 2t + B \cos 2t.$$

Substitution into the differential equation gives

$$-4A \sin 2t - 4B \cos 2t + 4A \sin 2t + 4B \cos 2t = \sin 2t,$$

which simplifies to $0 = \sin 2t$ so that something has gone wrong, since we cannot have $\sin 2t = 0$ for all t.

We can see what the problem is if we examine the complementary function for the differential equation. As we found in Example 8.26, the complementary function is $y_c = A \sin 2t + B \cos 2t$. This is the same function we tried above for y_p, so $y_c = y_p$. Since $y_c'' + 4y_c = 0$, we cannot expect it to also satisfy $y_c'' + 4y_c = \sin 2t$. It turns out that in this case, the correct form of the particular integral is

$$y_p = At \sin 2t + Bt \cos 2t.$$

Taking this as the particular integral we use the product rule to obtain

$$y_p'' = -4Bt \cos 2t - 4At \sin 2t - 4B \sin 2t + 4A \cos 2t.$$

Putting these results for y_p'' and y_p into $y'' + 4y = \sin 2t$ gives

$$-4B \sin 2t + 4A \cos 2t = \sin 2t,$$

so that $A = 0$ and $B = -\frac{1}{4}$. Hence the particular integral is

$$y_p = -\frac{1}{4} t \cos 2t.$$

□

The two forms of particular integral illustrated in these examples will suffice for most of our purposes. We will give you the rules for finding a particular integral for equations of the form

$$a_1 y'' + a_2 y' + a_3 y = c_1 \cos bt + c_2 \sin bt,$$

where one (but not both) of the constants c_1 or c_2 may be zero. Some other cases are dealt with in some of the later examples and exercises. This procedure is known as the *method of undetermined coefficients*.

- If $\cos bt$ and $\sin bt$ do not occur in the complementary function, then the particular integral will have the form

$$y_p = A \cos bt + B \sin bt.$$

- If $\cos bt$ and $\sin bt$ do occur in the complementary function, then the particular integral will have the form

$$y_p = At \cos bt + Bt \sin bt.$$

- Once we have chosen the form of the particular integral, we substitute it into the differential equation and solve for A and B.

EXAMPLE 8.29
Solve the initial value problem

$$y'' + 4y = \sin t, \quad y(0) = 1, \quad y'(0) = -1.$$

Solution. The complementary equation is

$$y'' + 4y = 0$$

and its solution is

$$y_c = C_1 \cos 2t + C_2 \sin 2t.$$

Since $\sin t$ is not in the complementary function, we take

$$y_p = A \cos t + B \sin t.$$

Hence

$$y_p' = -A \sin t + B \cos t$$
$$y_p'' = -A \cos t - B \sin t.$$

Substitution into the differential equation gives

$$-A \cos t - B \sin t + 4A \cos t + 4B \sin t = \sin t,$$

so that

$$3A \cos t + 3B \sin t = \sin t.$$

Thus, $A = 0$ and $B = \frac{1}{3}$ and a particular integral is

$$y_p = \tfrac{1}{3} \sin t.$$

This is the same as the particular integral we assumed in Example 8.26. The general solution is

$$y_c + y_p = C_1 \cos 2t + C_2 \sin 2t + \tfrac{1}{3} \sin t.$$

If we apply the initial conditions we find $C_1 = 1$ and $C_2 = -\frac{2}{3}$, so the solution of the initial value problem is

$$y = \cos 2t - \tfrac{2}{3} \sin 2t + \tfrac{1}{3} \sin t.$$

$\square$

We will now apply some of the principles above to a variety of oscillation problems with an oscillatory driving force. We will keep particular examples as simple as possible, but with problems of this type hand calculations can quickly become long and tedious. A package such as *Mathematica* can prove very useful. We will keep our equations consistent in that they will all start at equilibrium with initial velocity zero, that is $y(0) = 0$ and $y'(0) = 0$. The driving force will be $a \sin bt$ and it is this driving force that starts the motion and then maintains it.

8.6.2 The case of no friction

The initial value problem we are interested in is

$$y'' + 2\gamma y' + \omega^2 y = a \sin bt, \quad y(0) = 0, \quad y'(0) = 0 \qquad (8.19)$$

and we begin by assuming there is no friction, that is, the case $\gamma = 0$. This reduces to the initial value problem

$$y'' + \omega^2 y = a \sin bt, \quad y(0) = 0, \quad y'(0) = 0. \qquad (8.20)$$

Rather than considering the general case (8.20), we will consider particular illustrative examples.

EXAMPLE 8.30

We first consider the case $\omega \neq b$. This is the situation we considered in Example 8.29, namely

$$y'' + 4y = \sin t.$$

The general solution is

$$y = C_1 \cos 2t + C_2 \sin 2t + \tfrac{1}{3} \sin t.$$

Using the initial conditions $y(0) = 0, y'(0) = 0$ gives $C_1 = 0$ and $C_2 = -\tfrac{1}{6}$, so

$$y = -\tfrac{1}{6} \sin 2t + \tfrac{1}{3} \sin t.$$

The graph of the solution is shown in Figure 8.10. The solution is the sum of two oscillating functions,

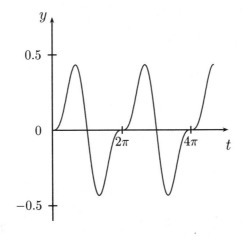

Figure 8.10: A periodic forced motion

one with period π and the other with period 2π.

Here the natural frequency of the system is $1/\pi$ while the frequency of the driving force is $1/(2\pi)$. Since one is a rational multiple of the other, the motion is periodic. This will not be the case if we had $y'' + 4y = \sin \pi t$, for example, where the ratio of the frequencies is irrational.

EXAMPLE 8.31

Next, consider the case $\omega = b$. An example of this case is the equation

$$y'' + 4y = \sin 2t, \quad y(0) = 0, \quad y'(0) = 0.$$

The complementary function
$$y_c = C_1 \cos 2t + C_2 \sin 2t$$
is the same as Example 8.30. Since the right hand side of the differential equation appears in the complementary function, the particular integral has the form
$$y_p = At \cos 2t + Bt \sin 2t.$$
Hence
$$y_p' = A \cos 2t - 2At \sin 2t + B \sin 2t + 2Bt \cos 2t$$
$$y_p'' = -4A \sin 2t - 4At \cos 2t + 4B \cos 2t - 4Bt \sin 2t.$$

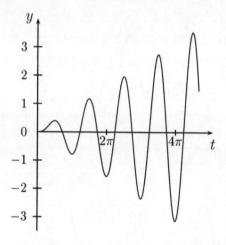

Figure 8.11: Resonance

Substituting into the differential equation gives
$$-4A \sin 2t + 4B \cos 2t = \sin 2t,$$
so that $A = -\frac{1}{4}, B = 0$ and the general soution is
$$y = C_1 \cos 2t + C_2 \sin 2t - \tfrac{1}{4} t \cos 2t.$$
Using the initial conditions gives $C_1 = 0, C_2 = \frac{1}{8}$ and the particular solution
$$y = \tfrac{1}{8} \sin 2t - \tfrac{1}{4} t \cos 2t.$$
As in Example 8.30, this solution is the sum of two oscillations. They both have the same frequency, but the second oscillation has amplitude increasing with t. In this case the oscillations increase without bound, a phenomenon known as *resonance* (Figure 8.11). In a real problem such unbounded oscillations cannot occur—either there is sufficient friction to prevent resonance occurring or the system fails because the oscillations become too large for it to bear.

The Tacoma Narrows Bridge mentioned in Chapter 1 is such a case. Another example is the loud squeals from amplifiers which can occur as a result of temporary resonance in the electrical circuits driving the system. To guard against resonance, it is important to know the natural frequencies of the system and the driving force.

EXAMPLE 8.32
Finally, we consider the case $\omega \approx b$. This leads to a different type of phenomenon, which occurs when the natural frequency of the system and the driving force are only slightly different. This situation gives rise to *beats* (Figure 8.12). It is difficult to find an easy hand calculation which demonstrates

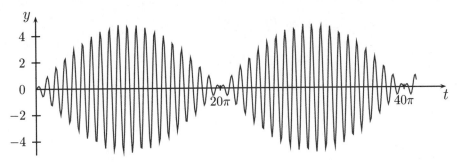

Figure 8.12: Beats

beats, but it is an ideal situation for a system such as *Mathematica*. We take

$$y'' + 4y = \cos 2.1t, \quad y(0) = 0, \quad y'(0) = 0.$$

A calculation similar to the ones above shows that

$$y = 2.44 \cos 2t - 2.44 \cos 2.1t.$$

Again this is made up of two oscillations, this time of slightly different frequencies. The nature of the solution can be made clearer if we rewrite it using the trigonometric identity

$$\cos(A+B) - \cos(A-B) = -2 \sin A \sin B.$$

If we put

$$A = \frac{2 + 2.1}{2} t$$

and

$$B = \frac{2 - 2.1}{2} t,$$

then

$$y = -4.88 \sin 0.05t \sin 2.05t.$$

The term $\sin 0.05t$ has a period of $2\pi/0.05 \approx 125$ while the term $\sin 2.05t$ has a period of $2\pi/2.05 \approx 3$. Thus the term $\sin 0.05t$ varies slowly compared with the term $\sin 2.05t$ and the motion is a rapid oscillation with frequency 2.05 Hz with a slowly varying sinusoidal amplitude of frequency 0.05 Hz. The graph is plotted in Figure 8.12.

8.6.3 The case where friction is present

This is the final case we consider in this chapter. There are obviously many variants depending on the strength of the damping and the frequency of the driving force. To see the general structure of the solutions we first analyse a case where damping is weak and there is a non-resonant driving force.

Consider the initial value problem

$$y'' + \frac{1}{4}y' + y = \sin 2t, \quad y(0) = 0, \quad y'(0) = 0$$

We find that

$$y_c = e^{-t/8}\left(C_1 \cos \frac{\sqrt{63}}{8}t + C_2 \sin \frac{\sqrt{63}}{8}t\right),$$

$$y_p = A \cos 2t + B \sin 2t.$$

Substituting y_p into the differential equation gives

$$\left(-3A + \frac{B}{2}\right)\cos 2t - \left(\frac{A}{2} + 3B\right)\sin 2t = \sin 2t.$$

Thus

$$-3A + \frac{B}{2} = 0$$

$$\frac{A}{2} + 3B = -1,$$

from which we find $A = -2/37$ and $B = -12/37$. Hence

$$y = y_c + y_p$$
$$= e^{-t/8}\left(A \cos \frac{\sqrt{63}}{8}t + B \sin \frac{\sqrt{63}}{8}t\right) - \frac{1}{37}(2\cos 2t + 12 \sin 2t).$$

Applying the initial conditions gives

$$C_1 = \frac{2}{37}, \quad C_2 = \frac{194}{37\sqrt{63}}$$

and so the solution is

$$y = \frac{2}{37}e^{-t/8}\left(\cos \frac{\sqrt{63}}{8}t + \frac{97}{\sqrt{63}} \sin \frac{\sqrt{63}}{8}t\right) - \frac{1}{37}(2\cos 2t + 12 \sin 2t).$$

This is a combination of oscillations. Since the term

$$\frac{2}{37}e^{-t/8}\left(\cos \frac{\sqrt{63}}{8}t + \frac{97}{\sqrt{63}} \sin \frac{\sqrt{63}}{8}t\right)$$

is damped by the $e^{-t/8}$ factor, it dies out as time increases. It is negligible after about 20 seconds. This part of the solution is called the *transient part* of the solution. The term

$$-\frac{1}{37}(2\cos 2t + 12\sin 2t)$$

is an oscillation at the driving frequency and dominates once the transient has died down. This is called the *steady state* part of the solution. The solution is plotted in Figure 8.13.

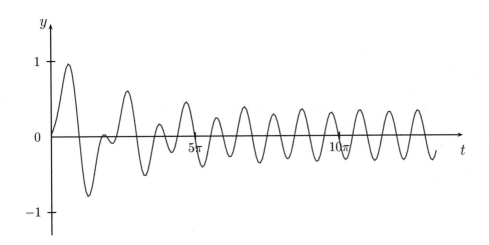

Figure 8.13: The solution to $y'' + \frac{1}{4}y' + y = \sin 2t$, $y(0) = 0$, $y'(0) = 0$

The next example illustrates a case where the damping is much stronger than in the case above. Let
$$y'' + 2y' + 4y = 13 \sin t, \quad y(0) = 0, \quad y'(0) = 0$$
be a given initial value problem. We have
$$y_c = e^{-t} \left(C_1 \cos \sqrt{3}t + C_2 \sin \sqrt{3}t \right),$$
$$y_p = A \cos t + B \sin t.$$
Substituting y_p into the differential equation gives
$$(2A + 2B) \cos t + (-2A + 3B) \sin t = 13 \sin t.$$
Thus
$$3A + 2B = 0$$
$$-2A + 3B = 13.$$
We solve these equations to get $A = -2$ and $B = 3$. Hence
$$y = y_c + y_p = e^{-t} \left(C_1 \cos \sqrt{3}t + C_2 \sin \sqrt{3}t \right) - 2 \cos t + 3 \sin t.$$
Applying the initial conditions gives $C_1 = 2$, $C_2 = -1/\sqrt{3}$ and so the solution is
$$y = y_c + y_p = e^{-t} \left(2 \cos \sqrt{3}t - \frac{1}{\sqrt{3}} \sin \sqrt{3}t \right) - 2 \cos t + 3 \sin t.$$

This is again a combination of a transient part and a steady state part, but now the damping factor e^{-t} is stronger than in the previous case and the transient part dies out much more rapidly. The solution is shown in Figure 8.14. The contribution of the transient part is only apparent during the first second and is very small.

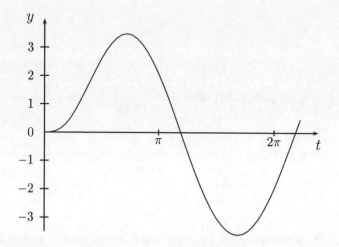

Figure 8.14: The solution to $y'' + 2y' + 4y = 13\sin t$, $y(0) = 0$, $y'(0) = 0$

This structure of transient and steady state solutions applies in any situation apart from very small damping when an irregular transient solution or resonance may persist.

8.6.4 More on undetermined coefficients

The method of undetermined coefficients (page 155) can be applied to the equation

$$ay'' + by' + cy = f(t)$$

in cases where $f(t)$ is not of the form $c_1 \cos bt + c_2 \sin bt$. We have to make an intelligent guess about the form of the particular integral y_p. This is illustrated in the worked examples below.

EXAMPLE 8.33
Solve the equation
$$y'' + 3y' + 2y = 4.$$

Here we take $y_p = a$, where a is a constant. We substitute this into the given equation to get

$$\text{LHS} = 0 + 0 + 2a$$
$$= \text{RHS}$$
$$= 4,$$

so that $y_p = a = 2$.

The complementary function is the solution of $y'' + 3y' + 2y = 0$ and is easily seen to be $y_c = Ae^{-2x} + Be^{-x}$, so that the general solution is

$$y = y_c + y_p$$
$$= Ae^{-2x} + Be^{-x} + 2.$$

EXAMPLE 8.34
Solve the equation
$$y'' + 3y' = 4.$$

The complementary function is the solution of $y'' + 3y' = 0$ and is easily seen to be $y_c = Ae^{-3x} + B$. If we attempt to take $y_p = a$ as we did in the above example, we will run into trouble, because this is just a special case of the complementary function. Instead, we take $y_p = ax$ and substitute this into the given differential equation to get

$$\text{LHS} = 0 + 3a$$
$$= \text{RHS}$$
$$= 4,$$

so $a = 4/3$ and $y_p = ax = 4x/3$. The general solution is

$$y = y_c + y_p$$
$$= Ae^{-3x} + B + \frac{4}{3}x.$$

EXAMPLE 8.35
Solve the equation
$$y'' + 3y' = 5e^{2x}.$$

The complementary function is the solution of $y'' + 3y' = 0$ and is easily seen to be $y_c = Ae^{-3x} + B$. We take $y_p = ae^{2x}$ and substitute this into the given differential equation to get

$$\text{LHS} = 4ae^{2x} + 6ae^{2x}$$
$$= \text{RHS}$$
$$= 5e^{2x},$$

so $a = 1/2$ and $y_p = ae^{2x} = e^{2x}/2$. The general solution is

$$y = y_c + y_p$$
$$= Ae^{-3x} + B + \frac{e^{2x}}{2}.$$

EXERCISES 8.6

Solve the following differential equations. Where initial conditions are given, find a particular solution satisfying the initial conditions. Otherwise, find the general solution.

1. $y'' + y' - 2y = 5\cos x$
2. $y'' + y = \sin 2x$

3. $y'' - y' = \sin x$

4. $y'' - y' + 9y = 3\sin 3x$

5. $y'' + 2y' + 5y = 3\sin 2x$

6. $x'' + x = \cos t$

7. $y'' + y = 4\sin x$, $\quad y(0) = 2$, $\quad y'(0) = 1$

8. $y'' + \omega^2 y = A\sin\omega t$, $\quad y(0) = 0$, $\quad y'(0) = 0$

9. $y'' + \omega^2 y = A\cos\lambda t$, $\quad y(0) = 0$, $\quad y'(0) = 0$, $\quad (\lambda \neq \omega)$

10. $x'' - 4x' + 3x = 17$

11. $x'' + x = 6e^t$

12. $x'' - 3x' + 2x = 6e^{3t}$

13. $x'' + 4x' = 8$

14. $x'' - x' - 6y = 2$

15. $x'' + 4x' + 2x = 5e^{-2t}$

16. The position $x(t)$ of a mass suspended from a spring satisfies the differential equation

$$x'' + 2\gamma x' + 4x = f(t).$$

(a) If $\gamma = 0$ and $f(t) = 5\sin at$, for what values of a does resonance occur?

(b) If $2\gamma = 5$ and $f(t) = 8$, find $\lim_{t\to\infty} x(t)$. What does this mean physically?

17. A mass of 30 kg is supported from a spring for which $k = 750$ N/m and is at rest. If a force of $20\sin 2t$ is now applied to the mass, find its position at any subsequent time t.

18. Give some examples of common phenomena in which resonance is important.

Electrical circuits can be modelled by linear differential equations. Consider the circuit below, which contains a time varying voltage source $E(t)$, an inductance L, a capacitance C and a resistance R. This is known as an *RLC circuit*.

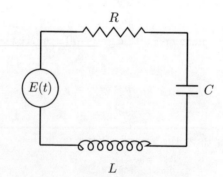

If $q(t)$ denotes the charge on the capacitor at time t, then the current $i(t) = dq/dt$. It can be shown that the charge q obeys the differential equation

$$L\frac{d^2q}{dt^2} + R\frac{dq}{dt} + \frac{1}{C}q = E(t).$$

If R, L and C are constant, then we can differentiate the above equation to get

$$L\frac{d^2i}{dt^2} + R\frac{di}{dt} + \frac{1}{C}i = E'(t).$$

19. Assume that in the above RLC circuit, the initial values of q and i are both zero. Find the charge and current at any time t if:

   (a) $R = 4$ ohms, $L = 1$ henry, $C = 1/13$ farads and $E(t) = 26$ volts. [Hint: In the equation for q, take $q_p = k$, where k is a constant]

   (b) $R = 10$ ohms, $L = 2$ henry, $C = 1/12$ farads and $E(t) = \sin 2t$ volts.

   (c) $R = 4$ ohms, $L = 1$ henry, $C = .5$ farads and $E(t) = \cos t$ volts.

20. An RLC circuit has $L = 0.2$ henry, $C = 8 \times 10^{-7}$ farad and $E(t) = 0$. Find the resistance R for which the circuit is critically damped.

CHAPTER 9
INTEGRATION

In Chapter 1 we began with three problems. These were formulated in terms of differential equations, which we were able to solve after developing suitable mathematical methods and concepts. However, we did not consider all aspects of these problems. For instance, in the *Tower of Terror*, we only dealt with the motion on the vertical part of the tower. If we want to analyse the motion around the curved part of the track (BC in Figure 1.2 of Chapter 1), the mathematics we have developed is inadequate to do this. In this chapter, we will begin to develop methods which can be used to solve this problem. These new methods will also provide an alternative way of solving the previous problems.

9.1 ANOTHER PROBLEM ON THE *TOWER OF TERROR*

Let us concentrate on the curved part of the motion in the *Tower of Terror* (Figure 9.1). We assume that the curved part of the track is an arc of a circle of radius r metres. We can specify the position of the car on this arc by using the angle θ shown in the diagram, where $0 \leqslant \theta \leqslant \pi/2$. This angle will be a function of t and the problem is to find this function, that is, to find the rule for calculating values of θ from values of t.

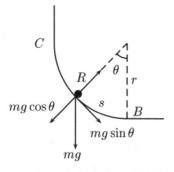

Figure 9.1: At the bottom of the *Tower of Terror*.

There are two forces on the car—gravity and the normal reaction R. We can resolve the force of gravity into a component $mg \sin \theta$ acting tangentially to the circle and a component $mg \cos \theta$ acting perpendicularly to the circle, where m is the mass of the car. The car has an acceleration as it traverses the circle. This comes about because it is being forced to move in a circle rather than a straight line and because its rate of traversing the circle may be changing. It can be shown that this acceleration can be resolved into a radial component $r\theta'^2$ and a tangential component $r\theta''$. Since the radius of the circular arc does not change, the frame must apply a reaction force $R = mr\theta'^2$ as well as a reaction to the radial component of its weight, that is $mg\cos\theta$. The total reaction force supplied by the frame is thus $mg\cos\theta + mr\theta'^2$, so that the net radial force on the car is $mr\theta'^2$. This net force keeps the car on the circle and prevents it flying off on a tangent, but does not affect its speed and plays no role in our problem. Consequently, we need only consider the tangential force $mg\sin\theta$.

Let s denote the distance along the curve from B, the lowest portion of the curve. Using Newton's second law of motion we have

$$m\frac{d^2s}{dt^2} = -mg\sin\theta.$$

If r is the radius of the circular arc and θ is measured in radians, then $s = r\theta$ and so

$$mr\frac{d^2\theta}{dt^2} = -mg\sin\theta$$

or

$$\theta'' + \omega^2 \sin\theta = 0,$$

where $\omega^2 = g/r$. This is the differential equation we have to solve to find θ as a function of t. If we were to try to solve this equation by assuming a power series for θ of the form

$$\theta = a_0 + a_1 t + a_2 t^2 + a_3 t^3 + \cdots$$

for θ, we would run into difficulty with the term

$$\sin\theta = \sin\left(a_0 + a_1 t + a_2 t^2 + a_3 t a^3 + \cdots\right).$$

The basic problem is that the differential equation is not linear and so our methods for linear equations are not applicable. Since infinite series cannot be easily used to solve this problem, we shall develop an alternative approach. Rather than leaping into the unknown by considering this problem now, it is easier to go back to the original three problems presented in Chapter 1 and look for alternative ways of solving them. We can then return to the present problem and use our new methods to solve it. This will take some time and we will not be able to complete the solution to the problem of finding the motion along the circular arc until Chapter 13.

9.2 MORE ON AIR PRESSURE

The initial value problem

$$\frac{dy}{dx} = y, \quad y(0) = 1$$

arose in our discussion of the air pressure problem in Chapter 6. We solved it by a power series method to get the solution $y = e^x$. We are now going to get this same result by a different method, although it will take us most of this chapter and the next to do this.

The initial condition we used was $y(0) = 1$, so for values of x close to zero, y is not zero and we can write the differential equation as

$$\frac{1}{y}\frac{dy}{dx} = 1. \tag{9.1}$$

The key to the new method is to use the chain rule on the left-hand side of this equation. Suppose that we are able to find a function F whose derivative is given by $F'(t) = 1/t$. Then equation (9.1) can be written

$$F'(y)\frac{dy}{dx} = 1,$$

or

$$\frac{d}{dx}\big(F(y(x))\big) = 1. \tag{9.2}$$

We now have a new problem: find a function whose derivative is 1. Such functions are easy to find. We have

$$\frac{d}{dx}(x) = 1$$
$$\frac{d}{dx}(x+2) = 1$$
$$\frac{d}{dx}(x+3\pi) = 1,$$

and, in general,

$$\frac{d}{dx}(x+c) = 1$$

for any constant c. As we shall later show, any function with derivative 1 must have the form $x + c$, so let's assume this for the moment. Then we can write equation (9.2) as

$$F\big(y(x)\big) = x + c, \tag{9.3}$$

that is,

$$(F \circ y)(x) = x + c.$$

At this stage we know the following:

- F is an unknown function whose derivative satisfies $F'(x) = 1/x$.

- The composition $F \circ y$ satisfies $(F \circ y)(x) = x + c$.

It is not obvious that we have achieved anything. The solution $y(x)$ that we want is tied up with the unknown function F in the composition $F \circ y$ and it is difficult to see how to disentangle it. However, we have identified two problems that need to be explored in greater detail.

- If we are given a function f, how do we find a function F such that $F' = f$? The answer to this question will lead us into *primitive functions* (or *antiderivatives*) and *integrals*, which we discuss in the next section.

- If we know the function F and the composition $F \circ y$, how do we find y? To answer this question we will need to understand *inverse functions*. These will be discussed in Chapter 10.

9.3 INTEGRALS AND PRIMITIVE FUNCTIONS

Let f be a given function. A function F such that $F' = f$ is called a *primitive*, an *antiderivative* or an *indefinite integral* of f. Finding F is the reverse of the problem of differentiation. For example, x is a primitive of 1, $\cos x$ is a primitive of $-\sin x$ and so on. There are no comprehensive rules for finding primitives and the process is intrinsically harder than differentiation.

The first point to notice about primitives is that they are not unique. A given function has many primitive functions. Thus x^2, $x^2 + 1$ and $x^2 + \pi + 44$ are all primitives of $2x$. Since the derivative of a constant is zero, it is easy to see that we can add a constant to any primitive function to get a new primitive function. In other words, if F is a primitive function of f, then so is $F + c$ for any constant c. In fact, we have the following result.

THEOREM 9.1
If F and G are primitives of the same function, then $F = G + c$ for some constant c.

This theorem is a consequence of the result (which we stated on page 100) that a function with zero derivative on an interval is constant on the interval. Suppose F and G are primitives of f. Then $F' = f, G' = f$ and so

$$\frac{d}{dx}(F(x) - G(x)) = F'(x) - G'(x) = f(x) - f(x) = 0.$$

Thus $F(x) - G(x) = c$, where c is a constant, and hence $F(x) = G(x) + c$.

A consequence of this theorem is that once we have found one primitive of a function, we have found all of them. The others differ by constants. There is a special notation used for primitives. If F is a primitive of f, we write

$$\int f(x)\,dx = F(x) + c.$$

This is read as *the (indefinite) integral of $f(x)$ with respect to x is $F(x)$ plus c*. Note that *Mathematica* does not put in the constant c. The notation has arisen for historical reasons and we shall have more to say about it later.

Knowing results for derivatives enables us to find primitives:

- $\int x^n\,dx = \dfrac{x^{n+1}}{n+1} + c, \quad n \neq -1$

- $\int \cos ax\,dx = \dfrac{1}{a}\sin ax + c$

- $\int \sin ax\,dx = -\dfrac{1}{a}\cos ax + c$

- $\int e^{ax}\,dx = \dfrac{1}{a}e^{ax} + c$

EXERCISES 9.3

Find primitives for the functions in Exercises 1–6.

1. $f(x) = x^3 + 2x + 1$
2. $f(x) = e^{4x} + \sin 2x$

3. $\phi(x) = \sec^2 x + \sin 3x$ 4. $g(x) = \dfrac{1}{x^3}$

5. $y = x^2(x-1)$ 6. $f(x) = (x-2)^2$

7. A function F satisfies $F'(t) = 3t^2 + 1$ and $F(0) = 3$. Determine F.

9.4 AREAS UNDER CURVES

In Section 9.1 we assumed it was possible to find a function F whose derivative satisfies $F'(x) = 1/x$. Can we in fact determine a convenient expression for F? Certainly the primitive of F is not a power function since

$$\int x^n \, dx = \frac{x^{n+1}}{n+1} + c, \quad n \neq -1,$$

where c is an arbitrary constant. It is possible that there is some complicated rational function which when differentiated by chain rules, product rules and the like may, after much cancelling, end up as $1/x$. This is unlikely. Usually the process of differentiation results in a more complicated function. In fact it can be shown (with some difficulty) that there is *no* function defined by an algebraic formula whose derivative is $1/x$. So is there any such function F at all? Around the period 1650–1670, Sir Isaac Newton and Gottfried Leibniz developed the ideas which led to one of the great discoveries in calculus. If we are given *any* continuous function f, there is a way of defining a function F such that $F' = f$. This is done by a process known as *definite integration* and provides a third way in which to define functions. Recall that we have so far defined functions in one of two ways:

- The simplest types of function were those defined by algebraic formulas such as

$$f(x) = \frac{x^2 + 3x + 1}{x^3 + 2}.$$

- Functions defined by power series arose when we solved differential equations, but in general any power series defines a function whose domain is the interval of convergence of the power series.

In this chapter we will show you how to use integration to produce functions having a given derivative.

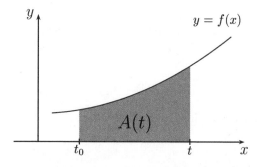

Figure 9.2: The area under a curve

Calculus is full of surprises. One of the most fundamental of these arises as follows. Let t_0 be a fixed point in the domain of a positive continuous function f (Figure 9.2). Consider the area $A(t)$ of the region bounded by the graph of f, the x axis, and the two vertical lines $x = t_0$ and $x = t$, where t is any point in the domain of f which satisfies $t > t_0$. We loosely refer to $A(t)$ as the *area under the curve* $y = f(x)$ *between* t_0 *and* t. Notice that in this way we can define a function (denoted by A) if we let the ouput number be $A(t)$ for the input number t. The surprising fact is that $A' = f$. Why should this be so? A function and its primitive are connected by differentiation, which can be physically interpreted as the slope of a curve at a given point, and this does not appear to have anything to do with the area under the curve. It is certainly not a result we might have expected. There is an even greater surprise in store. The connection between area and primitives is not just mathematical entertainment. It also turns out to be useful.

Before proceeding with the full details we shall consider a few simple cases to illustrate the connection between primitives and area.

EXAMPLE 9.1
Let $g(x) = x$. The region bounded by the x axis, the line $y = x$ and the vertical line through a point $x = t$ on the positive x axis is a triangle with base t and height t (Figure 9.3), so that its area is $A(t) = \frac{1}{2}t^2$. Here we have $A' = g$, so that A is a primitive of g.

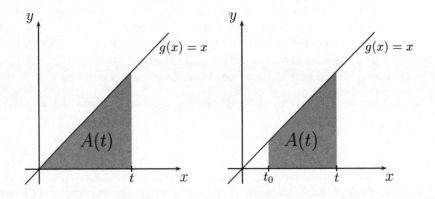

Figure 9.3: Regions under $g(x) = x$

EXAMPLE 9.2
Let us take the same function $g(x) = x$ as in the previous example, but this time we consider the area $A(t)$ of the region bounded by $y = x$, the x axis and the vertical lines through $x = t_0$ and $x = t$, $t \geqslant t_0$ (Figure 9.3). It is easy to see that

$$A(t) = \tfrac{1}{2}t^2 - \tfrac{1}{2}t_0^2,$$

and that $A' = g$. Hence A is again a primitive of g, but differs from the primitive in the previous example by a constant.

EXAMPLE 9.3
Let $g_1(x) = -x$. The region bounded by the x axis, the line $y = -x$ and the vertical line through a point $x = t$ on the positive x axis is a triangle with base t and height t (Figure 9.4), so that its area is $A(t) = \frac{1}{2}t^2$. In this case, it is $-A$ which is a primitive of g_1, that is $(-A)' = g_1$.

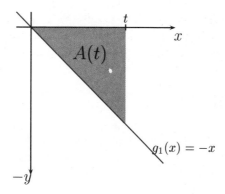

Figure 9.4: A triangular region under $g_1(x) = -x$

EXAMPLE 9.4

Let $h(x) = x^2$. We wish to find the area $A(t)$ of the region bounded by the x axis, the parabola $y = x^2$ and the vertical line through a point $x = t$ on the positive x axis (Figure 9.5). We can solve this problem by making use of a result which has been known for over 2000 years and which was first proved by Archimedes in about 250 BC[1]. Archimedes found that the area of the parabolic segment POQ in Figure 9.5 is $\frac{4}{3}at$. Using this result, we see that

$$A(t) = at - \tfrac{2}{3}at$$
$$= \tfrac{1}{3}at$$
$$= \tfrac{1}{3}t^3,$$

since $a = t^2$. We see that $A' = h$, so A is a primitive of h. □

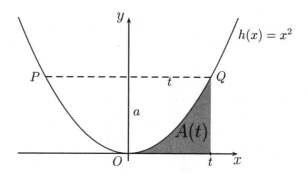

Figure 9.5: The area under a parabola

---

[1] This discovery is documented in a remarkable piece of work entitled *Quadrature of the parabola*. In it, Archimedes performs what is essentially a numerical integration to reach his goal of finding the area of a parabolic segment. The significance of his ideas were not realised until the seventeenth century when the development of the calculus occurred. Interested readers may like to read this work, which can be found in the *Encyclopaedia Britannica* series, *Great Books of the Western World* (volume 11).

The key observation to be made in all of these examples is that the area of the region between the graph of the given function and the x axis is, to within a sign, also a primitive for the function. This is not a coincidence; rather it is one of the most fundamental facts about the calculus and one which will need to be carefully investigated.

9.5 AREA FUNCTIONS

Let f be a given function and let t_0 be a fixed, but arbitrary, point on the x axis in the domain of f. We call t_0 a *reference point*. We want to define another function F by making use of the area of the region bounded by the graph of f, the x axis and the vertical lines through t_0 and any other point t in the domain of f, with $t > t_0$. We shall have to be careful about signs when doing this. We begin with the most straightforward case by assuming $f(x) > 0$ for all x in the domain of f. Then (Figure 9.6) we define $F(t)$ by

$$F(t) = A(t),$$

where $A(t)$ is the area of the region under the curve between t_0 and t.

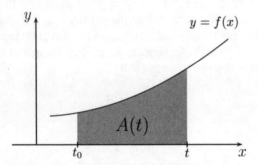

Figure 9.6: Defining a function F by $F(t) = A(t)$

The area $A(t)$ (and hence the function value $F(t)$) is a number which can be calculated for each choice of t. The calculation is not always easy and we shall have to investigate methods for doing it, but for the moment let's accept that it can be done.

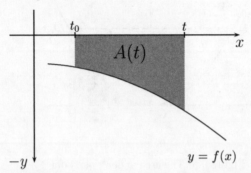

Figure 9.7: Defining a function F by $F(t) = -A(t)$

Now let us take the case when $f(x) < 0$ for all x in the domain of f (Figure 9.7). As before, choose a reference point t_0 and let t be any point in the domain of f satisfying $t > t_0$. Let $A(t)$ be the

area of the region bounded by the x axis, the graph of f and the vertical lines through t_0 and t. Then we define $F(t)$ by
$$F(t) = -A(t).$$

Finally, we take the case where $f(x)$ changes sign one or more times for $t_0 < x < t$. In this case, some parts of the region bounded by the lines $x = t_0$, $x = t$, the graph of f and the x axis will lie above the x axis, while other parts will lie below the x axis. We attach a positive sign to the area of each part of the region which lies above the x axis and a negative sign to the area of each part of the region which lies below the x axis. The function F is defined by letting its output number $F(t)$ at the point t be the sum of these signed areas (Figure 9.8).

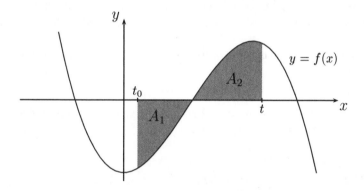

Figure 9.8: Defining a function F by $F(t) == -A_1 + A_2$

In summary we can use the area of the region bounded by the graph of a function f, the x axis and two vertical lines to define a function F. In Examples 9.1–9.4 this function F satisfies the condition $F' = f$ (Exercises 9.5). There are two matters which we need to address:

- How do we calculate the area of the region under a curve in the general case? We need to be able to do this in order to calculate function values of F.

- Does the function F always satisfy $F' = f$?

The answers to these questions emerged with the discovery of the calculus by Newton and Leibniz in the middle of the 17th century. The answer to the first question is conceptually straightforward, although the calculations may sometimes become unwieldy. We shall address this question in the next two sections. The answer to the second question is given by the fundamental theorem of the calculus, which we shall discuss in Section 9.8.

EXERCISES 9.5

1. Construct the function F described in the text of this section for each of the examples 9.1–9.4. Verify that $F' = f$ in each case.

2. Use Archimedes' result for the area of a parabolic segment to show that the area of the region bounded by the graph of $f(x) = 3x^2$, the x axis and the vertical lines $x = 2$ and $x = t$ is given by $A(t) = t^3 - 8$. Is A a primitive of f?

9.6 INTEGRATION

In the previous sections, we have shown how it is possible to construct a function F by using the area (with an appropriate sign) of the region bounded by the graph of a continuous function f, the x axis and the lines $x = t_0$ and $x = t$, where t_0 and t are in the domain of f. The first problem to be dealt with is to determine a method of calculating the area of a figure with one or more curved boundaries. The basic idea is to approximate the area by simple shapes whose area can be calculated and then take a limit. This idea is a very ancient one. It was used by Archimedes to calculate the area of the parabolic segment we considered earlier.

9.6.1 Positive functions

Let f be a function defined on the interval $[a, b]$ and suppose that $f(x) > 0$ for all $x \in [a, b]$. We wish to determine the area of the region bounded by $y = f(x)$, the x axis and the lines $x = a$ and $x = b$ (Figure 9.9).

The simplest way to approximate the area of the region is by rectangles. Divide the interval $[a, b]$ into n subintervals

$$[x_0, x_1], [x_1, x_2], \ldots, [x_{n-1}, x_n],$$

where $x_0 = a$ and $x_n = b$. The intervals need not be of the same length. In each subinterval $[x_{i-1}, x_i]$ ($i = 1, 2, \ldots, n$), select an arbitrary point x_i^* and form the sum

$$\begin{aligned} S(n) &= f(x_1^*)(x_1 - x_0) + f(x_2^*)(x_2 - x_1) + f(x_3^*)(x_3 - x_2) + \cdots + f(x_n^*)(x_{n-1} - x_n) \\ &= f(x_1^*)\Delta x_1 + f(x_2^*)\Delta x_2 + f(x_3^*)\Delta x_3 + \cdots + f(x_n^*)\Delta x_n \\ &= \sum_{i=1}^{n} f(x_i^*)\Delta x_i, \end{aligned}$$

where $\Delta x_i = x_i - x_{i-1}$, $i = 1, 2, \ldots, n$. Such a sum is called a *Riemann sum*. It is the sum of

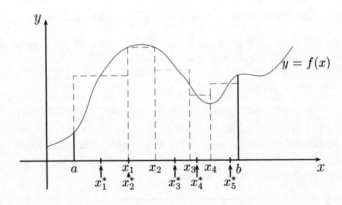

Figure 9.9: Approximating the integral by a Riemann sum

the areas of the individual rectangles shown in Figure 9.9. Such sums approximate the area under the curve and it seems intuitively clear that the approximation converges to the exact value of the area as we take more intervals in such a way that their number increases without limit and such that they all get arbitrarily small. This is true if f is continuous. However, it is quite difficult to prove this, since we have to consider all the possible ways in which the intervals become small and increase in number. We shall omit the detail on how to do it in a mathematically correct way and simply write

$$\int_a^b f(x)\,dx = \lim_{\substack{\Delta x_i \to 0 \\ n \to \infty}} \left(\sum_{i=1}^n f(x_i^*)(x_i - x_{i-1}) \right). \tag{9.4}$$

The expression on the left-hand side is called the *definite integral*[2] of f on the interval $[a, b]$. The numbers a and b are called the *limits of integration*. We call a the lower limit and b the upper limit.

The process of using Riemann sums to compute integrals can be simplified considerably if we take intervals of equal width. Begin by dividing the interval $[a, b]$ into n subintervals

$$[x_0, x_1], [x_1, x_2], \ldots, [x_{n-1}, x_n],$$

of width

$$\Delta x = \frac{b-a}{n},$$

where $x_0 = a$ and $x_n = b$.

There are now two obvious ways to approximate the area by rectangles. Firstly, we can use rectangles of height $f(x_{i-1})$ and base Δx, $(i = 1, 2, \ldots, n)$. This gives the following approximation to the area under the curve

$$\begin{aligned} S_L(n) &= f(x_0)\Delta x + f(x_1)\Delta x + \cdots + f(x_{n-1})\Delta x \\ &= \frac{b-a}{n}\Big(f(x_0) + f(x_1) + \cdots + f(x_{n-1})\Big). \end{aligned} \tag{9.5}$$

The particular Riemann sum $S_L(n)$ is called a *left-hand sum*, because we have used the value of the function at the left-end of each interval. This approximation is illustrated in Figure 9.10.

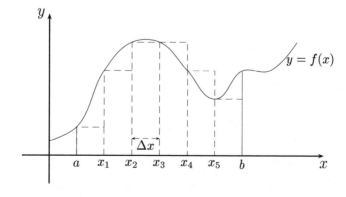

Figure 9.10: The integral approximated by a left-hand sum

Alternatively, we can calculate a *right-hand sum* by using the rectangles of height $f(x_i)$ and base

---

[2] The integral sign $\int$ was first used by Leibniz on October 29, 1675. It was an elongated "S", for the word "sum".

Δx, $(i = 1, 2, \ldots, n)$, as illustrated in Figure 9.11. This gives the approximation

$$S_R(n) = f(x_1)\Delta x + f(x_2)\Delta x + \cdots + f(x_n)\Delta x$$
$$= \frac{b-a}{n}\Big(f(x_1) + f(x_2) + \cdots + f(x_n)\Big). \tag{9.6}$$

It seems intuitively clear that as we allow n to increase, both of the sequences $\{S_L(n)\}$ and $\{S_R(n)\}$

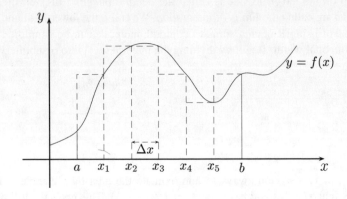

Figure 9.11: The integral approximated by a right-hand sum

will converge to a number which is equal to the area of the region in question and equal to the limit of the Riemann sum (9.4). Thus for any continuous positive function f, the integral

$$\int_a^b f(x)\,dx$$

may be calculated by taking a limit of left-hand sums:

$$\int_a^b f(x)\,dx = \lim_{n\to\infty} \frac{b-a}{n}\Big(f(x_0) + f(x_1) + \cdots + f(x_{n-1})\Big) \tag{9.7}$$

or by taking a limit of right-hand sums:

$$\int_a^b f(x)\,dx = \lim_{n\to\infty} \frac{b-a}{n}\Big(f(x_1) + f(x_2) + \cdots + f(x_n)\Big). \tag{9.8}$$

9.6.2 General functions

If f is not necessarily positive on the interval $[a, b]$, we can still divide $[a, b]$ into n subintervals

$$[x_0, x_1], [x_1, x_2], \ldots, [x_{n-1}, x_n],$$

and construct the limit of Riemann sums

$$\lim_{\substack{\Delta x_i \to 0 \\ n \to \infty}} \left(\sum_{i=1}^n f(x_i^*)(x_i - x_{i-1})\right),$$

where x_i^* is an arbitrary point in the i'th subinterval. As long as the function f is continuous on $[a, b]$, it can be shown that such a sum will converge and once again we denote the value of this sum by

$$\int_a^b f(x)\,dx.$$

As before, we call this the *definite integral of f* on the interval $[a, b]$. However, the interpretation of the integral is now different. In this case, some parts of the region bounded by the lines $x = a$, $x = b$, the x axis and the graph of f may lie above the x axis, while other parts may lie below it. We attach a positive sign to the area of each part of the region which lies above the x axis and a negative sign to the area of each part of the region which lies below the x axis. The integral is then the sum of these areas, taking the sign into account. This statement follows directly from the way in which the Riemann sum was constructed. If, for example, $f(x) < 0$ for all $x \in [a, b]$, then the limit of the sum in equation (9.4) will be negative.

There are some minor points to clear up. In the above discussion, we have assumed that $b > a$ in the integral $\int_a^b f(x)\,dx$. If this is not the case, we simply *define* $\int_a^b f(x)\,dx$ for $b < a$ by

$$\int_a^b f(x)\,dx = -\int_b^a f(x)\,dx.$$

Also, for equal upper and lower limits, we define

$$\int_a^a f(x)\,dx = 0.$$

9.6.3 Some properties of integrals

There are some other properties of integrals which we now state. First, there are the results for sums and multiples:

$$\int_a^b \left(f(x) + g(x)\right) dx = \int_a^b f(x)\,dx + \int_a^b g(x)\,dx$$

and

$$\int_a^b cf(x)\,dx = c\int_a^b f(x)\,dx,$$

where c is any constant.

These results can be proved in each case by expressing both sides either as the limit of a left-hand sum or as the limit of a right-hand sum and showing that these limits are equal.

There are no results for integrals of products and quotients which apply in all cases.

If we take a point c with $a < c < b$, then

$$\int_a^b f(x)\,dx = \int_a^c f(x)\,dx + \int_c^b f(x)\,dx.$$

If f is positive, then this result amounts to the addition of areas, but the same result holds whatever the order of a, b and c and whatever the sign of f. The use of left-hand or right-hand sums is inadequate to prove this result. A correct proof would need to use more general Riemann sums.

We can now return to the area functions we were discussing in Section 9.5. For any continuous function f with domain $[a, b]$ we can define a function F by

$$F(t) = \int_a^t f(x)\,dx, \quad a \leqslant t \leqslant b.$$

The discussion in Section 9.5 shows that the number $F(t)$ can be interpreted physically in terms of the area of a region bounded by the the graph of f and the vertical lines $x = a$ and $x = t$. Functions defined in this way will play a major role in the remainder of this book and it is important that you understand this process.

In general, integrals cannot be calculated exactly except in special cases. This is because no formula can be found for the Riemann sum and so there is no way of calculating the limit. We can compute approximate values to any required degree of accuracy by choosing sufficiently large values of n. These calculations are usually very tedious to do by hand and a computer is needed.

EXAMPLE 9.5
Evaluate

$$\int_1^2 \frac{1}{x}\,dx$$

approximately.

Here $a = 1$, $b = 2$. We choose $n = 10$, so that the width of the subintervals is $h = \frac{2-1}{10} = 0.1$. Then

$$\int_1^2 \frac{1}{x}\,dx \approx 0.1\left(\frac{1}{1.0} + \frac{1}{1.1} + \frac{1}{1.2} + \cdots + \frac{1}{1.9}\right) \doteq 0.719.$$

If we choose $n = 20$ ($h = 0.05$) we obtain 0.706 for the approximation while if we choose $n = 100$ ($h = 0.01$) we obtain 0.696. The sequence of values appears to be converging, but very slowly. There are various ways to improve the approximation and we shall deal with these in the next section. $\square$

In some special cases, it is possible to use formulas such as

$$1 + 2 + 3 + \cdots + n = \sum_{k=1}^{n} k = \tfrac{1}{2}n(n+1) \tag{9.9}$$

$$1^2 + 2^2 + 3^2 + \cdots + n^2 = \sum_{k=1}^{n} k^2 = \tfrac{1}{6}n(n+1)(2n+1) \tag{9.10}$$

$$1^3 + 2^3 + 3^3 + \cdots + n^3 = \sum_{k=1}^{n} k^3 = \tfrac{1}{4}n^2(n+1)^2 \tag{9.11}$$

to evaluate integrals explicitly by finding an expression for a left-hand or right-hand sum and then taking a limit. The above formulas can be proved by the use of mathematical induction.

EXAMPLE 9.6
Evaluate

$$\int_0^1 f(x)\,dx,$$

where $f(x) = x^2$.

Solution. We divide the interval $[0,1]$ into n subintervals of width $1/n$. Then using equation (9.8) with $a = 0, b = 1, x_1 = 1/n, x_2 = 2/n, \ldots x_n = n/n$, we find

$$\int_0^1 x^2 \, dx = \lim_{n \to \infty} \left[\frac{1}{n} \left(f\left(\frac{1}{n}\right) + f\left(\frac{2}{n}\right) + \cdots + f\left(\frac{n}{n}\right) \right) \right]$$

$$= \lim_{n \to \infty} \left[\frac{1}{n} \left(\frac{1}{n^2} + \frac{2^2}{n^2} + \frac{3^2}{n^2} + \cdots + \frac{n^2}{n^2} \right) \right]$$

$$= \lim_{n \to \infty} \left[\frac{1}{n^3} (1^2 + 2^2 + 3^2 + \cdots + n^2) \right]$$

$$= \lim_{n \to \infty} \left[\frac{1}{n^3} \times \frac{n(n+1)(2n+1)}{6} \right]$$

$$= \frac{1}{3}.$$

Thus

$$\int_0^1 x^2 \, dx = \frac{1}{3}.$$

Note that this gives Archimedes's formula quoted earlier.

EXERCISES 9.6

Evaluate the sums in Exercises 1–9.

1. $\sum_{k=1}^{10} (3k - 1)$

2. $\sum_{n=1}^{12} (n^2 + 3)$

3. $\sum_{k=1}^{12} (k - 1)^2$

4. $\sum_{p=1}^{8} (p+1)(p-2)$

5. $\sum_{i=1}^{6} (i^3 - 2i^2)$

6. $\sum_{k=4}^{k=10} (k^2 + 3k)$

7. $\sum_{k=5}^{20} k^3$

8. $\sum_{k=1}^{n} (k - n)^2$

9. $\sum_{k=1}^{p} (k^2 + 1)$

10. Determine the limits of the following sequences.

   (a) $\left\{ \dfrac{1 + 2 + 3 + \cdots + n}{n^2} \right\}$

   (b) $\left\{ \dfrac{1^2 + 2^2 + 3^2 + \cdots + n^2}{n^3} \right\}$

   (c) $\left\{ \dfrac{1^3 + 2^3 + 3^3 + \cdots + n^3}{n^4} \right\}$

In each of the following exercises, find the area of the region under the graph of f from $x = a$ to $x = b$ by expressing the integral as the limit of a sum.

11. $f(x) = 3x + 1, \quad a = 0, \quad b = 4$

12. $f(x) = 3x^2$ $a = 0$, $b = 5$

13. $f(x) = 3x^2$ $a = 1$, $b = 5$

14. $f(x) = 3x^2 + 2x + 1$ $a = 0$, $b = 5$

15. $f(x) = 10 - x^2$ $a = 1$, $b = 4$

16. $f(x) = (x - 1)^2$ $a = 0$, $b = 6$

9.7 EVALUATION OF INTEGRALS

In this section we shall assume for simplicity that all functions are positive. The formulas we get will hold even when this restriction is lifted, but we shall not prove it here.

9.7.1 The trapezoidal rule

It would appear from a consideration of Figure 9.12 that a better approximation to the area under a curve would be obtained if we used trapeziums instead of rectangles. Trapeziums are four-sided figures with a pair of parallel sides. To get an approximation to $\int_b^a f(x)\,dx$ we divide the interval $[a, b]$ into n subintervals of width $h = (b - a)/n$ and approximate the region under the curve between $x = a$ and $x = b$ by trapeziums, as shown in Figure 9.12. The area of a trapezium is easily derived by dividing it into two triangles and a rectangle. In the notation of the right-hand diagram in Figure 9.12, the area is found to be $\frac{1}{2}h(a + b)$. Using this result, we see that the area of a typical trapezium in Figure 9.12 is $\frac{1}{2}h\bigl(f(x_{i-1}) + f(x_i)\bigr)$, where $1 \leqslant i \leqslant n$.

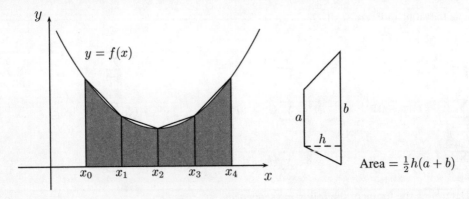

Figure 9.12: Approximating a region by trapeziums

Let T_n denote the sum of the areas of all such trapeziums. Then T_n is an approximation to $\int_a^b f(x)\,dx$. We find that

$$T_n = \frac{h}{2}\bigl(f(x_0) + f(x_1)\bigr) + \frac{h}{2}\bigl(f(x_1) + f(x_2)\bigr) + \cdots + \frac{h}{2}\bigl(f(x_{n-1}) + f(x_n)\bigr)$$

$$= \frac{h}{2}\bigl(f(x_0) + 2f(x_1) + 2f(x_2) + \cdots + 2f(x_{n-1}) + f(x_n)\bigr) \qquad (9.12)$$

Equation (9.12) is known as the *trapezoidal rule*.

EXAMPLE 9.7
Approximate
$$\int_1^2 \frac{1}{x}\,dx$$
using the trapezoidal rule for the cases $n = 10$, $n = 20$ and $n = 100$.

Solution. For $n = 10$, we have
$$T_{10} = \frac{0.1}{2}\left(\frac{1}{1.0} + \frac{2}{1.1} + \frac{2}{1.2} + \cdots + \frac{2}{1.9} + \frac{1}{2.0}\right) \doteq 0.6938.$$

For $n = 20$, we find
$$T_{20} \doteq 0.6933,$$
while for $n = 100$
$$T_{100} \doteq 0.6932.$$

The convergence is far more rapid than for the rectangle approximation.

9.7.2 Simpson's rule

An even better approximation is to take points three at a time and fit a parabola through them. To approximate
$$\int_a^b f(x)\,dx$$
we divide the interval $[a, b]$ into an even number of subintervals. Let
$$x_0 = a, x_1, x_2, \ldots, x_{n-1}, x_n = b$$
be the end points of these subintervals, where n is an even integer.

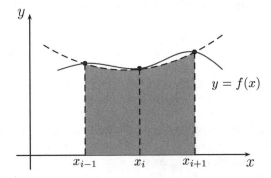

Figure 9.13: A parabolic approximation to part of the graph of $y = f(x)$.

For a typical triplet $\{x_{i-1}, x_i, x_{i+1}\}$ ($i = 1, 2, \ldots, n-1$), we pass a parabola through the points $(x_{i-1}, f(x_{i-1}))$, $(x_i, f(x_i))$ and $(x_{i+1}, f(x_{i+1}))$ and then find the area of the region bounded by this parabola, the x axis and the lines $x = x_{i-1}$ and $x = x_{i+1}$ (Figure 9.13). We can use the result of Archimedes mentioned earlier to do this. Denote the sum of these areas by S_n. Then S_n is an approximation to the integral $\int_a^b f(x)\,dx$ and we find, after some effort, that

$$S_n = \frac{h}{3}\Big(f(x_0) + 4f(x_1) + 2f(x_2) + 4f(x_3) + \cdots$$
$$\cdots + 2f(x_{n-2}) + 4f(x_{n-1}) + f(x_n)\Big). \tag{9.13}$$

Equation (9.13) is known as *Simpson's rule*.[3] It is sufficiently accurate for many practical purposes, but there are times when more sophisticated methods are needed. We shall not discuss these here. By choosing successively smaller values of h we look for a sequence of constant output numbers from the application of Simpson's rule.

EXAMPLE 9.8
Take $n = 10$, $n = 20$ and $n = 100$ in Simpson's rule to approximate the integral

$$\int_1^2 \frac{1}{x}\,dx.$$

We have
$$S_{10} = \frac{0.1}{3}\left(\frac{1}{1} + \frac{4}{1.1} + \frac{2}{1.2} + \frac{4}{1.3} + \cdots + \frac{4}{1.9} + \frac{1}{2}\right) \doteq 0.693150.$$

For $n = 20$ we find
$$S_{20} \doteq 0.693147$$

and for $n = 100$
$$S_{100} \doteq 0.693147.$$

We conclude that the value of the integral is 0.6931 to 4 decimal places.

In using approximations such as the trapezoidal rule or Simpson's rule, questions regarding the accuracy of these rules naturally arise. Let f be a function with domain $[a, b]$. Let T_n and S_n respectively denote the trapezoidal estimate and Simpson's estimate for $\int_a^b f(x)\,dx$. We can write

$$\int_a^b f(x)\,dx = T_n + \varepsilon(T_n)$$

and, for even values of n,

$$\int_a^b f(x)\,dx = S_n + \varepsilon(S_n).$$

Thus $|\varepsilon(T_n)|$ is the numerical difference between the actual value of the integral and the trapezoidal estimate with n subintervals, while $|\varepsilon(S_n)|$ is the numerical difference between the actual value of the integral and Simpson's estimate with n subintervals.

We state the following two results without proof.

- Suppose that f'' is continuous on $[a, b]$ and that there is a constant K such that $|f''(x)| \leqslant K$ for all x in $[a, b]$. Then

$$|\varepsilon(T_n)| \leqslant \frac{K(b-a)^3}{12n^2}.$$

---

[3]Thomas Simpson (1710–1761) was a weaver who taught himself mathematics. The rule which we know by his name was known to Gregory and Cavalieri in the seventeenth century, but Simpson popularised it in his book *A new treatise on fluxions*.

- Suppose that $f^{(4)}$ is continuous on $[a, b]$ and that there is a constant M such that $|f^{(4)}(x)| \leqslant M$ for all x in $[a, b]$. Then, if n is even

$$|\varepsilon(S_n)| \leqslant \frac{M(b-a)^5}{180n^4}.$$

EXAMPLE 9.9

Estimate the error in applying the trapezoidal rule with 10 subintervals to the integral

$$\int_1^2 \frac{1}{x}\,dx.$$

Solution. We put $f(x) = 1/x$, so that

$$|f''(x)| = \left|\frac{2}{x^3}\right|$$
$$\leqslant 2$$

on the interval $[1, 2]$. Consequently, using the above error estimate with $K = 2$, $a = 1$ and $b = 2$ gives

$$|\varepsilon(T_n)| \leqslant \frac{2(2-1)^3}{12 \times 10^2}$$
$$\doteq 0.0017.$$

Hence, using the result from Example 9.7, we have

$$\int_1^2 \frac{1}{x}\,dx = 0.6938 \pm 0.0017.$$

EXERCISES 9.7

In the problems below, calculate both the trapezoidal approximation T_n and Simpson's approximation S_n to the given integral. Use the indicated number of subintervals and round the answers to 4 decimal places.

1. $\int_1^3 x^2\,dx, \quad n = 4$
2. $\int_0^1 \sin x\,dx, \quad n = 6$
3. $\int_0^1 \frac{1}{1+x^2}\,dx, \quad n = 4$
4. $\int_0^1 \sqrt{1+x}\,dx, \quad n = 4$

The next two exercises give tabulated values for a function f. Calculate an approximation to

$$\int_a^b f(x)\,dx$$

using (a) the trapezoidal rule and (b) Simpson's rule.

5. $a = 1, b = 2.50$

x	1.00	1.25	1.50	1.75	2.00	2.25	2.50
$f(x)$	1.41	1.50	1.58	1.66	1.73	1.80	1.87

6. $a = 0, b = 10$

x	0	1	2	3	4	5	6	7	8	9	10
$f(x)$	0	-7	-24	-45	-64	-75	-72	-49	0	81	200

7. Two students are standing on a bridge over a creek. They decide to measure the volume of water flowing under the bridge in the following way. First, they measure the velocity of the water by throwing twigs into the water and timing how long it takes for them to pass under the bridge. They find the average velocity to be 0.5 m/sec. Next, by using a long rod, they measure the depth of water at various points along the width of the creek and derive the profile shown in the diagram, where all depths are given in metres. The creek is 8 metres wide. Use Simpson's rule to calculate the volume of water per second flowing under the bridge.

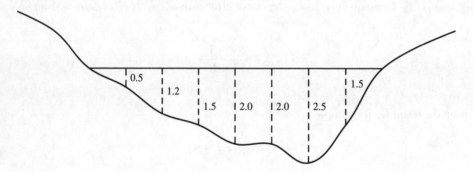

8. A function f is given by the rule

$$f(x) = \int_1^x \sqrt{1 + \sin t}\, dt.$$

Use Simpson's rule to evaluate $f(3)$.

9. Determine how large n must be in order to guarantee that the trapezoidal estimate T_n for the integral

$$\int_1^2 \frac{1}{x}\, dx$$

differs from its true value by no more that 0.0005.

10. Determine how large n must be in order to guarantee that Simpson's estimate S_n for the integral

$$\int_1^2 \frac{1}{x}\, dx$$

differs from its true value by no more that 10^{-6}.

9.8 THE FUNDAMENTAL THEOREM OF THE CALCULUS

We are now in a position to be able to solve the following problem raised in Section 9.2:

Given a function f, find a function F such that $F' = f$.

The following theorem provides the answer.

> **THEOREM 9.2 The fundamental theorem of the calculus**
> *If f is continuous on the interval $[a, b]$ and if F is defined by*
> $$F(x) = \int_a^x f(t)\,dt,$$
> *for $x \in [a, b]$ then $F'(x) = f(x)$.*

The function F is well defined because we can calculate its function values by procedures such as Simpson's rule. The theorem is proved using differentiation by first principles. We will only give an outline of the proof without going into detail. Take a sequence $\{h_n\}$ converging to zero. By definition

$$F'(x) = \lim_{n \to \infty} \frac{1}{h_n}\bigl(F(x + h_n) - F(x)\bigr).$$

Using differentiation by first principles, we consider

$$\frac{1}{h_n}\bigl(F(x + h_n) - F(x)\bigr) = \frac{1}{h_n}\left(\int_a^{x+h_n} f(t)\,dt - \int_a^x f(t)\,dt\right)$$

$$= \frac{1}{h_n}\left(\int_x^{x+h_n} f(t)\,dt\right)$$

Now if h_n is very small, then this integral is approximately $h_n \times f(x)$. Thus

$$\frac{1}{h_n}\bigl(F(x + h_n) - F(x)\bigr) \approx \frac{1}{h_n} h_n f(x) = f(x).$$

Taking the limit carefully as $n \to \infty$ (and this is where we gloss over the detail) we find that the approximation becomes an equality, that is

$$\lim_{n \to \infty} \frac{1}{h_n}\bigl(F(x + h_n) - F(x)\bigr) = f(x).$$

Hence
$$F'(x) = f(x).$$

Another way of writing this result is

$$\frac{d}{dx} \int_a^x f(t)\,dt = f(x).$$

EXAMPLE 9.10
Find a function F such that

$$F'(x) = \frac{1}{x}.$$

188 INTEGRATION

The function we are seeking follows directly from the fundamental theorem of the calculus. It is

$$F(x) = \int_a^x \frac{1}{t}\,dt,$$

where a is any positive constant and x is any number in $(0, \infty)$.

EXERCISES 9.8

Find the derivative of each of the functions in Exercises 1–8.

1. $f(x) = \int_1^x (1+t^2)\,dt$

2. $h(x) = \int_0^x (1+\sin 2t)\,dt$

3. $f(x) = \int_3^x \left(t - \frac{1}{t^2}\right) dt$

4. $f(t) = \int_1^t (x + e^{3x})\,dx$

5. $f(x) = \int_0^{x^2} (2t+1)\,dt$

6. $g(u) = \int_1^{2u^2} \sin t^2\,dt$

7. $\phi(x) = \int_a^{1+x^2} (1-u^2)^2\,du$

8. $f(x) = \int_1^{x^3} \alpha(t)\,dt$

9.9 THE LOGARITHM FUNCTION

The above example gives us a lead-in to the logarithm function. Let us define a function $\log x$ with domain $(0, \infty)$ by

$$\log x = \int_1^x \frac{1}{t}\,dt.$$

This function is often denoted by $\ln x$. The function value $\log x$ is equal to the area of the region under the curve $y = 1/t$ between the vertical lines $t = 1$ and $t = x$ (Figure 9.14) and is calculated by, for example, Simpson's rule. It follows from the definition and the fundamental theorem of the calculus that

$$\frac{d}{dx}(\log x) = \frac{1}{x}. \tag{9.14}$$

EXAMPLE 9.11
In Example 9.8, we found

$$\int_1^2 \frac{1}{x}\,dx \doteq 0.6931$$

to 4 decimal places. We conclude that $\log 2 \doteq 0.6931$.

Similar calculations give

$$\log \tfrac{1}{2} = \int_1^{\frac{1}{2}} \frac{1}{t}\,dt$$

$$= -\int_{\frac{1}{2}}^1 \frac{1}{t}\,dt$$

$$\doteq -0.6931.$$

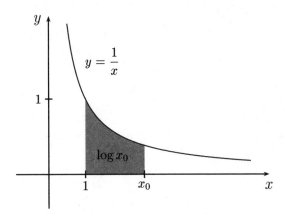

Figure 9.14: The logarithm function

and
$$\log 3.5 = \int_1^{3.5} \frac{1}{t}\, dt$$
$$\doteq 1.2527.$$

□

Let's now return to the matter of solving the initial value problem
$$\frac{dy}{dx} = y, \quad y(0) = 1.$$

In Section 9.2 we showed that by choosing a function F such that $F'(x) = 1/x$, we could write the above differential equation as
$$\frac{d}{dx}\bigl(F(y(x))\bigr) = 1, \tag{9.15}$$
from which we found
$$F(y(x)) = x + c.$$
We now know that $F(x) = \log x$, so we can write
$$\log(y(x)) = x + c.$$
To get the constant c we use the initial value $y(0) = 1$ to get $\log 1 = 0 + c$. It follows from the definition of $\log x$ that $\log 1 = 0$, so that $c = 0$. The solution to the initial value problem is $\log y(x) = x$.

The only problem that remains is to write the unknown function y explicitly in terms of x, that is, to extract the $y(x)$ from inside the logarithm. In order to do this, we will need to derive some properties of the logarithm function. In the next chapter we will discuss the matter of writing y explicitly in terms of x.

The graph of $\log x$ crosses the x axis at the point $x = 1$. In fact, it follows from the definition of the logarithm in terms of integrals that

$$\log x > 0 \text{ if } x > 1,$$
$$\log x < 0 \text{ if } 0 < x < 1,$$
$$\log 1 = 0.$$

Since the domain of $\log x$ is $(0, \infty)$, the graph does not cut the y axis.

We can also find the concavity of the graph of $\log x$. Since

$$\frac{d}{dx}(\log x) = \frac{1}{x}, \quad \frac{d^2}{dx^2}(\log x) = -\frac{1}{x^2}$$

we have

$$\frac{d}{dx}(\log x) > 0, \quad \frac{d^2}{dx^2}(\log x) < 0$$

for all $x > 0$, that is, for all x in the domain of $\log x$. Thus the graph is increasing and concave down for all $x > 0$. As x gets close to zero, the slope of the graph increases and approaches infinity. It is not yet clear what happens as x becomes arbitrarily large. Does the slope of the graph approach a number, or does it become arbitrarily large? A software package such as *Mathematica* has the logarithm function built in, and we can use it to plot the graph in Figure 9.15. In principle you could plot such a graph by using Simpson's rule to calculate $\log x$ for different x values, plotting these points and then joining them with a smooth curve. In fact, this is what *Mathematica* essentially does.

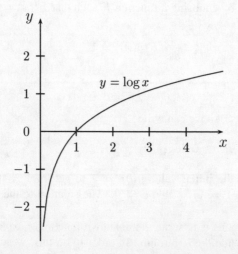

Figure 9.15: Graph of $y = \log x$

THEOREM 9.3 Properties of log x
For all positive numbers a and b and all rational numbers r, we have

$$\log ab = \log a + \log b$$
$$\log \frac{1}{a} = -\log a$$
$$\log \frac{a}{b} = \log a - \log b$$
$$\log a^r = r \log a.$$

To prove this, consider the function $F(x) = \log ax - \log x$. Using the chain rule we have

$$F'(x) = \frac{d}{dx}(\log ax - \log x) = \frac{1}{ax}a - \frac{1}{x} = 0$$

Since $\log ax - \log x$ has derivative zero, we have $\log ax - \log x = c$ for some constant c. To find c, put $x = 1$. Then $c = \log a$, so $\log ax - \log x = \log a$. Finally, put $x = b$ and rearrange the last equation to get

$$\log ab = \log a + \log b.$$

Next we have

$$\log \frac{1}{a} + \log a = \log \frac{1}{a}a$$
$$= \log 1$$
$$= 0,$$

so that

$$\log \frac{1}{a} = -\log a.$$

The next identity in Theorem 9.3 now follows easily:

$$\log \frac{a}{b} = \log a \frac{1}{b}$$
$$= \log a + \log \frac{1}{b}$$
$$= \log a - \log b.$$

The final identity of Theorem 9.3 is proved in a number of stages. Here we will only prove the result if r is an integer. If r is a positive integer n we have

$$\log a^n = \log(\underbrace{a \times a \times \cdots \times a}_{n \text{ terms}})$$
$$= \underbrace{\log a + \cdots + \log a}_{n \text{ terms}}$$
$$= n \log a.$$

Next, if r is a negative integer n, we can write $n = -m$, where m is a positive integer. Then

$$n \log a = -m \log a$$
$$= m \log \frac{1}{a}$$
$$= \log \left(\frac{1}{a}\right)^m$$
$$= \log a^{-m}$$
$$= \log a^n.$$

We have shown that

$$n \log a = \log a^n,$$

where n is any non-zero integer. That the result is also true if $n = 0$ is left as an exercise.

We shall show on page 205 that this result holds for all rational numbers r. It is also true if r is any real number, but we will not go into detail on this matter.

Questions about the behaviour of $\log x$ for large and small values of x can now be answered. Since $\log 2 = 0.693$, we have $\log 2^n = 0.693 \times n$, so $\log x$ gets arbitrarily large as x increases. This is usually written

$$\lim_{x \to \infty} \log x = \infty.$$

Similarly, $\log 2^{-n} = -0.693 \times n$, so $\log x$ gets arbitrarily small as x gets closer to zero. We write this as

$$\lim_{x \to 0^+} \log x = -\infty.$$

The notation $x \to 0^+$ is used to emphasize that x can only approach zero through positive values.

9.9.1 The indefinite integral of $1/x$

We have shown that

$$\frac{d}{dx}(\log x) = \frac{1}{x}.$$

In this equation, we must have $x > 0$ in order for $\log x$ to be defined. Thus

$$\int \frac{1}{x} dx = \log x + c, \quad x > 0. \tag{9.16}$$

If $x < 0$, we put $u = -x$, so that $u > 0$. Then

$$\frac{d}{dx}(\log(-x)) = \frac{d}{du}(\log u) \frac{du}{dx}$$
$$= \frac{1}{u}(-1)$$
$$= \frac{1}{x}.$$

Hence

$$\int \frac{1}{x} dx = \log(-x) + c, \quad x < 0. \tag{9.17}$$

We can combine equations (9.16) and (9.17) into the single equation

$$\int \frac{1}{x}\, dx = \log|x| + c, \quad x \neq 0. \tag{9.18}$$

The graph of $y = \log|x|$ is shown in Figure 9.16.

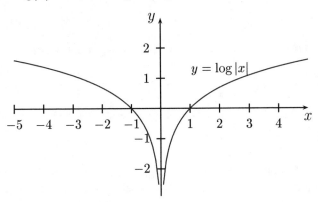

Figure 9.16: The graph of $y = \log|x|$

9.9.2 Logarithmic differentiation

Logarithmic differentiation is a technique for simplifying the process of finding a derivative in certain cases. If we are given $f(x)$ and want to find $f'(x)$, it may be easier to begin by finding the derivative of $\log f(x)$. The process is best illustrated by example.

EXAMPLE 9.12
Find $f'(x)$ if
$$f(x) = \frac{(x+1)(x+2)}{(x+3)(x+4)}.$$

Solution. We take logarithms to get

$$\log f(x) = \log\left[\frac{(x+1)(x+2)}{(x+3)(x+4)}\right]$$
$$= \log(x+1) + \log(x+2) - \log(x+3) - \log(x+4).$$

Differentiating both sides of this equation gives

$$\frac{f'(x)}{f(x)} = \frac{1}{x+1} + \frac{1}{x+2} - \frac{1}{x+3} - \frac{1}{x+4}.$$

Hence

$$f'(x) = f(x)\left[\frac{1}{x+1} + \frac{1}{x+2} - \frac{1}{x+3} - \frac{1}{x+4}\right]$$
$$= \frac{(x+1)(x+2)}{(x+3)(x+4)}\left[\frac{1}{x+1} + \frac{1}{x+2} - \frac{1}{x+3} - \frac{1}{x+4}\right].$$

EXERCISES 9.9

Differentiate the functions given in Exercises 1–11.

1. $f(x) = \log(2x + 4)$
2. $f(x) = \log(x^2 + 1)$
3. $f(x) = \log(1 + x)^4$
4. $f(x) = \log(\ln(x))$
5. $f(x) = x^2 \log(x^2 + 1)$
6. $f(x) = \log(x\sqrt{1 + x^2})$
7. $f(x) = \log \dfrac{1+x}{1-x}$
8. $f(x) = \log \sqrt{\dfrac{1+x^2}{1-x^2}}$
9. $f(x) = \log(x^2 + 1)(1 + x^3)$
10. $f(x) = \log(1 + 2x)^3(x^2 - 1)^2$
11. $f(x) = \log(x + e^{-x})$

Logarithms have a number of applications. For physiological reasons, scales of luminosity and sound are often measured logarithmically. Suppose the eye is presented with a series of lamps of absolute brightness one, two, three, four, ... units, all at the same distance, and then with another series of lamps of brightness one, two, four, eight, ... units, then it is the second series and not the first which will appear to form a uniform gradation of brightness. Similar conclusions apply to sounds. These considerations are the reason we adopt the *stellar magnitude scale* for measuring the brightness of stars and other celestial objects and the *decibel* scale for the measurement of sound.

12. In 200BC, the Greek astromomer Hipparchus, wishing to describe the brightness of stars, classified them in terms of *magnitudes*. The brightest stars were said to be of first magnitude. The next group of prominent, but less bright stars were classified as being of the second magnitude, and so on, down to the stars which are just visible to the naked eye, which were classified as being of the sixth magnitude.

    In the nineteenth century, when accurate measurements became possible, it was found that a sixth magnitude star was almost exactly 100 times fainter than one of the first magnitude. In fact, the following table of relative brightnesses was constructed.

Magnitude	1	2	3	4	5	6
Brightness	100	40	16	6.3	2.5	1

    (a) By what factor does the brightness of a star change for a unit change in magnitude?

    (b) Plot a graph of the logarithm of the brightness against the magnitude. What type of graph do you get?

    (c) Show that if B is the brightness of a star, scaled to 100 for first magnitude stars, and m is its magnitude, then
    $$\ln B \doteq 5.53 - 0.92m.$$

13. The stellar magnitude scale described in the previous example can be extended at both ends. A seventh magnitude star is 2.512 times fainter than a sixth magnitude star. It would only be visible in a telescope. Similarly, a star of magnitude zero is 2.512 times brighter than one of the first magnitude and a star of magnitude -1 is 2.512 times brighter again. On this scale the sun is of magnitude -26.7.

(a) Show that if a star of magnitude m_1 has brightness B_1 and a star of magnitude m_2 has brightness B_2, then
$$\ln \frac{B_1}{B_2} \doteq 0.92(m_2 - m_1).$$

(b) Let B_s denote the brightness of the sun and let B_f denote the brightness of the faintest star visible to the naked eye. Compute $\ln(B_s/B_f)$. By trial and error with your calculator, determine B_s/B_f. What is the physical meaning of this ratio?

14. In land animals it is found that the leg diameter is (not surprisingly) related to the body mass. Leg thickness increases in proportion to the body mass rather than to the linear dimensions of the animal, so that large animals, such as elephants, have relatively thicker legs than smaller animals, such as dogs. Here is a table giving the leg bone circumference (in centimeters) and body mass (in kilograms) for a selection of animals.

	Mouse	Cat	Human	Horse	Hippo	Elephant
Leg circumference	0.6	4.5	13.6	27.4	40	80.5
Body mass	0.05	3.5	50	350	1700	6000

(a) Plot the logarithm of the leg circumference on the vertical axis and the logarithm of the body mass on the horizontal axis. You should find that relationship between the variables is approximately linear. Use *Mathematica* to find the best straight line through the points you have plotted.

(b) What do you notice about the plotted point for the hippopotomus? Suggest a reason for your observation.

(c) A fossil leg bone of a *Brachiosaurus* was found to have a circumference of 1.33 metres. Estimate the mass of the animal when it was alive. The *Brachiosaurus* was one of the largest land animals of all time, although some incomplete fossil skeletons suggest that larger animals may have existed.

15. The range of human hearing is impressive. If the intensity of sound is measured in watts per square metre, a normal person can just detect a noise of 10^{-12} W/m^2. Permanent ear damage would occur at 100 W/m^2. As mentioned above, the ear perceives sounds logarithmically and this has given rise to the *decibel* scale. If S_0 is the intensity of a reference sound and S is a measured sound, then the difference between S and S_0 in decibels is given by

$$D = \frac{10}{\ln 10} \ln\left(\frac{S}{S_0}\right).$$

(a) Take $S_0 = 10^{-12}$ W/m^2 as the reference point on the decibel scale. The noise caused by a jet aircraft take-off at 60 m is 130 db. Compute the intensity of this sound in watts per square metre.

(b) A pneumatic riveter has a noise level of 10 W/m^2. Express this noise level in decibels.

CHAPTER 10
INVERSE FUNCTIONS

In solving the initial value problem
$$y' = y, \quad y(0) = 1$$
we obtained $\log y(x) = x$. The remaining problem we have is to extract y from this equation. This is done using *inverse functions*.

Let's start with some simple cases which have a similar pattern to this one. We want to solve problems of the following type:

- A function F is given.
- We know $F(y(x)) = x$.
- How do we find y?

EXAMPLE 10.1
Suppose $F(x) = 2x + 3$. Then if $F(y(x)) = x$, we have
$$2y(x) + 3 = x$$
and so
$$y(x) = \tfrac{1}{2}(x - 3).$$
We let $G(x) = \tfrac{1}{2}(x - 3)$. Then $F(y(x)) = x$ gives $y(x) = G(x)$.

EXAMPLE 10.2
Let $F(x) = x - 4$ and suppose that $F(y(x)) = x$. Then
$$y(x) - 4 = x$$
and so
$$y(x) = x + 4.$$
We let $G(x) = x + 4$. Then $F(y(x)) = x$ gives $y(x) = G(x)$. □

What is the relationship between the functions F and G in these examples? We can answer this question by applying the two functions in succession. Using F and G from Example 10.1 we have

$$F(G(x)) = F(\tfrac{1}{2}(x-3)) = 2 \times \tfrac{1}{2}(x-3) + 3 = x$$
$$G(F(x)) = G(2x+3) = \tfrac{1}{2}((2x+3) - 3) = x.$$

These relationships are illustrated in the diagram below.

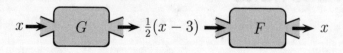

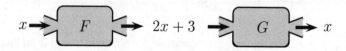

Figure 10.1: The action of a function and its inverse

Similarly, using F and G from Example 10.2 gives

$$F(G(x)) = F(x+4) = x + 4 - 4 = x$$
$$G(F(x)) = G(x-4) = x - 4 + 4 = x.$$

In both these cases each function has the effect of reversing the action of the other. If we think of a function as a set of operations which starts with an input number, operates on it and produces an output number then, in the above examples, the other function reverses these operations and produces the number we started from. Functions which have this property are called *inverse functions*; G is the inverse of F and F is the inverse of G. We normally write F^{-1} instead of G for the inverse of F. A function F and its inverse F^{-1} are related by the pair of equations

$$F(F^{-1}(x)) = x,$$
$$F^{-1}(F(x)) = x.$$

This is illustrated in terms of function machines in Figure 10.2. We will postpone a more precise definition of inverse functions to the next section.

The notation $F^{-1}(x)$ for the inverse function is universally used. Unfortunately, there is the danger of confusing it with the reciprocal of $F(x)$, that is, with $1/F(x)$. You should note carefully that the -1 in the notation for the inverse is not an exponent. The symbol $F^{-1}(x)$ *does not mean* $1/F(x)$. If we do want to express $1/F(x)$ in exponent form we will write $[F(x)]^{-1}$.

EXAMPLE 10.3
Let $f(x) = 3x + 4$. Find the inverse f^{-1} of f.

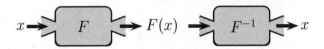

Figure 10.2: The action of a function and its inverse

Solution. We put
$$f(f^{-1}(x)) = x,$$
so that
$$3f^{-1}(x) + 4 = x.$$
Hence
$$f^{-1}(x) = \tfrac{1}{3}(x - 4).$$
Alternatively, we let $y = f^{-1}(x)$. Then
$$f(y) = x,$$
so that
$$3y + 4 = x.$$
Hence
$$y = \tfrac{1}{3}(x - 4)$$
and, as before,
$$f^{-1}(x) = \tfrac{1}{3}(x - 4).$$
□

Now we can solve the problem posed at the beginning of this chapter. If we know that a function F has an inverse F^{-1} and if
$$F(y(x)) = x,$$

then, applying F^{-1} to both sides of this equation, gives us

$$F^{-1}(F(y(x))) = F^{-1}(x).$$

Hence, since F^{-1} undoes the action of F,

$$y(x) = F^{-1}(x).$$

The initial value problem

$$y' = y, \quad y(0) = 1$$

thus has solution

$$y(x) = \log^{-1}(x).$$

This is all very well, but we now have two more problems to deal with. (Will this never end?)

- How can we be sure that a given function has an inverse?

- How can we calculate function values for the inverse, assuming it exists?

10.1 THE EXISTENCE OF INVERSES

Consider the function f defined by $f(t) = 2t + 3$. If we pick any output number y and put $f(t) = y$, then there is only one input number t which will give this output number. For example, if $y = 3$, then $f(t) = 3$ implies that $t = 0$. On the other hand, if we take the function $g(t) = t^2$, then there are two input numbers for each nonzero output number. For example, if $y = 4$, then $t = \pm 2$. If g were to have an inverse g^{-1}, then this inverse would satisfy $g^{-1}(g(2)) = 2$ and $g^{-1}(g(-2)) = -2$. This would require $g^{-1}(4) = 2$ and $g^{-1}(4) = -2$, which is impossible, because functions must give a unique output number for each input number. Hence in this case there can be no such function g^{-1} and the function g does not have an inverse. These arguments can be generalised to show that a function can have an inverse only if each output number comes from a unique input number. A function with this property is said to be *one-to-one*. We write 1–1 rather than one-to-one.

There is a graphical interpretation of 1–1 functions. Any function has the property that an input number gives only one output number, so that any vertical line will intersect the graph of the function in at most one point. A 1–1 function has the additional property that any *horizontal* line intersects the graph at most once. In Figure 10.3, the graph on the right is not 1–1, while the graph on the left is 1–1. You are asked to show in the Exercises that a function which is either increasing on an interval I or decreasing on an interval I will be 1–1 on I. Hence if the derivative of a function has constant sign on an interval I, then the function will be 1–1 on I. This test cannot be applied if the function is not differentiable at one or more points and other methods are needed.

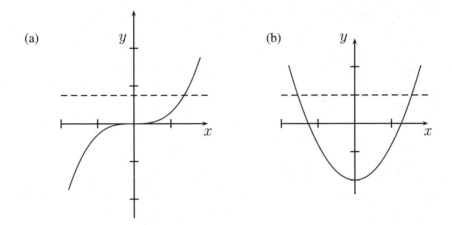

Figure 10.3: (a) A 1–1 function. (b) A function which is not 1–1.

We are almost ready to give a formal definition of an inverse function. First, we state the following result.

THEOREM 10.1 *Let f be a 1–1 function. Then there is a unique function g whose domain is the range of f and which satisfies the equation*

$$f(g(x)) = x$$

for all x in the range of f.

The proof is straightforward. If x is in the range of f, then it must come from some input number in the domain of f. Since f is 1–1, there can be only one such input number and we define $g(x)$ to be this number.

The function we have called g in the theorem is called the *inverse* of f and is usually denoted by f^{-1}.

DEFINITION 10.1 *Let f be a 1–1 function. The **inverse** of f, denoted f^{-1}, is the unique function whose domain is the range of f and which satisfies the equation*

$$f(f^{-1}(x)) = x$$

for all x in the range of f.

EXAMPLE 10.4

We show that the function $f(x) = 4x + 1$ is 1–1 and hence has an inverse. Choose any two real numbers x_1 and x_2. The function is 1–1 if distinct input numbers never give the same output number. This means that if $f(x_1) = f(x_2)$, then *necessarily* $x_1 = x_2$. We equate $f(x_1)$ and $f(x_2)$ and if this forces us to conclude that $x_1 = x_2$, then f is 1–1.

The equation $f(x_1) = f(x_2)$ implies that $4x_1 + 1 = 4x_2 + 1$ and hence that $x_1 = x_2$. Thus the function is 1–1. Alternatively, we see that $f'(x) = 4 > 0$, so that, by the remark on page 200, f is 1–1.

We leave it as an exercise to show that the inverse of f is given by $f^{-1}(x) = \frac{1}{4}(x - 1)$. □

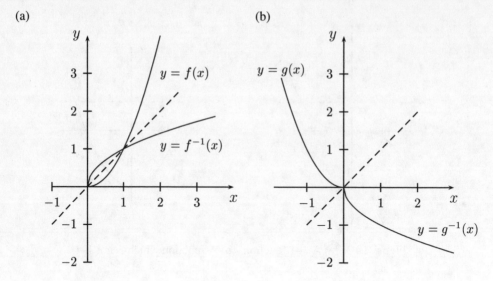

Figure 10.4: Graphs of (a) f and its inverse (b) g and its inverse.

There are cases where a function does not have an inverse, but where it would nevertheless be very convenient if it did. Often we can deal with this situation by restricting the domain of the original function. For example, the function

$$F(x) = x^2, \quad x \in \mathbb{R}$$

does not have an inverse, but both the functions

$$f(x) = x^2, \quad x \geqslant 0$$

and

$$g(x) = x^2, \quad x \leqslant 0$$

do have unique inverses. In fact, the function f and its inverse f^{-1} enable us to give a precise definition of square roots. To do this, let x be any positive number. Then

$$f(f^{-1}(x)) = x$$

so that

$$\left(f^{-1}(x)\right)^2 = x.$$

It is the number $f^{-1}(x)$ which is called the *square root* of x. The fact that inverses are unique guarantees that x has a unique positive square root, normally written $\sqrt{x}$. The inverse of f is $f^{-1}(x) = \sqrt{x}$ and the inverse of g is $g^{-1}(x) = -\sqrt{x}$. The graphs of these functions are shown in Figure 10.4.

> **DEFINITION 10.2 The square root**
> Let $a \geqslant 0$. Then the square root of a is the unique positive number t such that $t^2 = a$. We denote the square root of a by $\sqrt{a}$.

In a similar way, we can define nth roots of any positive number. The *n*th root of a is the unique positive number t such that $t^n = a$. We denote the nth root of a by $a^{1/n}$ or $\sqrt[n]{a}$.

10.1.1 The graphs of f and f^{-1}

There is an important connection between the graphs of f and f^{-1}. The graph of f consists of the set of all points of the form $(x, f(x))$. If we take $f(x)$ as the input number for f^{-1}, we get $f^{-1}(f(x)) = x$ as the output number. Hence the graph of f^{-1} consists of all points of the form $(f(x), x)$. This means that the two graphs will be symmetric with respect to the line $y = x$. Figure 10.4 shows this symmetry in two particular cases discussed earlier. Notice also that the range of f will be the domain of f^{-1} and the range of f^{-1} will be the domain of f.

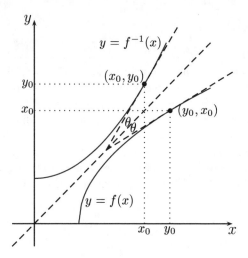

Figure 10.5: The slopes of f and f^{-1}

The relationship between the slopes of the graphs of f and f^{-1} is also a simple one. Since the graphs are mirror images of each other in the line $y = x$, it follows that the tangent line through $(x, f(x))$ on the graph of f is the mirror image of the tangent line through $(f(x), x)$ on the graph of f^{-1}. With the aid of trigonometry, it can be shown that the slopes of these two tangent lines are reciprocals of each other. Let m_1 be the slope of the tangent line to the graph of f^{-1} at the point (x_0, y_0) and let m_2 be the slope of the tangent line to the graph of f at the point (y_0, x_0) (Figure 10.5). Then

$$m_1 = \tan(\pi/4 + \theta)$$
$$= \frac{1 + \tan\theta}{1 - \tan\theta},$$

while

$$m_2 = \tan(\pi/4 - \theta)$$
$$= \frac{1 - \tan\theta}{1 + \tan\theta}.$$

Thus

$$m_1 = 1/m_2.$$

Since $m_1 = (f^{-1})'(x_0)$ and $m_2 = f'(y_0)$, we have

$$(f^{-1})'(x_0) = \frac{1}{f'(y_0)},$$

where $y_0 = f^{-1}(x_0)$. This relationship between the derivatives of a function and its inverse is summarised in the next theorem.

THEOREM 10.2 The derivative of an inverse function
Let f be a function having an inverse function f^{-1}. If f is differentiable at $f^{-1}(x)$ and if $f'(f^{-1}(x)) \neq 0$, then f^{-1} is differentiable at x and

$$(f^{-1})'(x) = \frac{1}{f'(f^{-1}(x))}.$$

You should try to understand this theorem with the aid of Figure 10.5. The value of $(f^{-1})'(x_0)$ is the slope m_1 of the tangent at (x_0, y_0) to the graph of f^{-1}. The theorem states that if the slope of the tangent at the point (y_0, x_0) to the graph of f is m_2, then $m_1 = 1/m_2$. This is a rather difficult theorem to come to grips with. You will need to be thoroughly familiar with function notation and you will probably have to think carefully about it before it begins to make sense.

Our main use of Theorem 10.2 will be to derive formulas for the derivatives of various functions of interest. As an illustration, we will use it to calculate the derivative of $\sqrt{x}$.

Let $f(x) = x^2$ with domain $x > 0$. Then $f'(x) = 2x$. The inverse function is given by $f^{-1}(x) = \sqrt{x}$. We have

$$\begin{aligned}
\frac{d}{dx}(\sqrt{x}) &= \frac{d}{dx}(f^{-1}(x)) \\
&= (f^{-1})'(x) \\
&= \frac{1}{f'(f^{-1}(x))} \\
&= \frac{1}{f'(\sqrt{x})} \\
&= \frac{1}{2\sqrt{x}},
\end{aligned}$$

so we have shown that

$$\frac{d}{dx}(\sqrt{x}) = \frac{1}{2\sqrt{x}}. \tag{10.1}$$

The reasoning we have applied to square roots can be applied to nth roots for any positive integer n. We can show that for $x > 0$

$$\frac{d}{dx}(x^{1/n}) = \frac{1}{n}x^{(1/n)-1}. \tag{10.2}$$

If n is odd, then this formula also holds for $x < 0$.

The chain rule, together with equation (10.2) can be used to prove the following important result.

THEOREM 10.3 The derivative of x^n
For all rational numbers r, we have

$$\frac{d}{dx}(x^r) = rx^{r-1}. \tag{10.3}$$

The result holds on every open interval on which x^n is defined.

We conclude this section by showing that
$$n \log x = \log x^n$$
for all rational numbers n. This result was first stated in Theorem 9.3 of Chapter 9, but was only proved for integer values of n. We have
$$\frac{d}{dx}(\log x^n - n \log x) = \frac{nx^{n-1}}{x^n} - \frac{n}{x}$$
$$= \frac{n}{x} - \frac{n}{x}$$
$$= 0.$$

Hence
$$\log x^n - n \log x = c,$$
where c is a constant. Putting $x = 1$ shows that $c = 0$. Hence $\log x^n = n \log x$.

EXERCISES 10.1

Determine whether or not the given function is one-to-one and, where possible, find the inverse function.

1. $f(x) = 2x + 3$
2. $f(x) = x^2 - 1$
3. $f(x) = x^3 + x$
4. $f(x) = x + \frac{1}{x}$
5. $f(x) = \frac{1}{x^3 + 1}$
6. $f(x) = \frac{x+1}{x-3}$
7. $f(x) = |x|$
8. $f(x) = 5x + 2$
9. $f(x) = x^3 - 1$
10. $f(x) = (1-x)^3$
11. $f(x) = \frac{x}{1+x}$

12. Let f be a function which is either increasing on the interval $[a, b]$ or decreasing on the interval $[a, b]$. Show that f is 1–1 on $[a, b]$.

13. Show that the function $f(x) = x^n$, $x > 0$, has a unique inverse $f^{-1}(x) = x^{1/n}$. Use this result together with Theorem 10.2 to derive equation (10.2).

14. Use the chain rule and equation (10.2) to derive equation (10.3).

10.2 CALCULATING FUNCTION VALUES FOR INVERSES

In simple cases we can use algebraic manipulation to find an exact expression for the inverse of a given function. The idea is to solve the equation $f(f^{-1}(x)) = x$ for $f^{-1}(x)$. You may prefer to write y for $f^{-1}(x)$. It is easier to work with the symbol y rather than the string $f^{-1}(x)$.

EXAMPLE 10.5
Show that the function
$$f(x) = \frac{1}{x^2 + 1}, \quad x > 0$$

is 1–1 and find its inverse f^{-1}.

Solution. Suppose $f(x_1) = f(x_2)$. Then we have
$$\frac{1}{x_1^2 + 1} = \frac{1}{x_2^2 + 1},$$
so that
$$x_2^2 + 1 = x_1^2 + 1.$$
Hence $x_2^2 = x_1^2$, and therefore $x_1 = x_2$, since both of them are positive. It follows that f is 1–1.

To find f^{-1}, let $y = f^{-1}(x)$ and solve $f(y) = x$. We get
$$\frac{1}{y^2 + 1} = x,$$
so that
$$y = \sqrt{\frac{1}{x} - 1},$$
and hence
$$f^{-1}(x) = \sqrt{\frac{1}{x} - 1}.$$

□

In examples such as the one above we were able to find a formula for the inverse function without a lot of difficulty. However, this is not usually the case. How, for example, can we find the inverse of more complicated functions such as the logarithm function? We need a method of calculating values of $f^{-1}(x)$ and one way of doing this is to use *Newton's method*.

10.2.1 Newton's method

The idea behind the method of finding roots of equations known as *Newton's method* was discovered by Sir Isaac Newton around 1660. The form in which it is used today is not precisely the one discovered by Newton. Newton's technique was modified by Joseph Raphson in 1690 and by Thomas Simpson (of Simpson's rule fame) in 1740. The method is often called the *Newton–Raphson method*. Nevertheless, the essential idea is Newton's.

Suppose we want to solve the equation $f(x) = 0$. We can usually arrive at an approximate solution x_0 by various means—plot the graph of $f(x)$ and estimate the root or plug in values of x into the formula for $f(x)$ until we get one that gives $f(x) \approx 0$. Let x_s denote the exact solution we are seeking. Figure 10.6 shows the graph of $y = f(x)$.

The equation of the tangent at $(x_0, f(x_0))$ is
$$y - f(x_0) = f'(x_0)(x - x_0).$$
This intersects the x axis at
$$x_1 = x_0 - \frac{f(x_0)}{f'(x_0)}$$
and x_1 will (usually) be a better approximation to x_s than x_0. We repeat this process by taking x_1 as the new starting value. In general, if we have an approximation x_n to the root, then the next approximation x_{n+1} will be
$$x_{n+1} = x_n - \frac{f(x_n)}{f'(x_n)}.$$

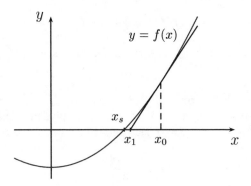

Figure 10.6: Newton's method

We thus generate a sequence of values $\{x_n\}$, which in most cases converges to the root x_0.

EXAMPLE 10.6
Let's solve the equation $x^3 - 2 = 0$. Since x will be a cube root of 2, we expect it to lie between 1 and 2. Newton's rule for this case is

$$x_{n+1} = x_n - \frac{x_n^3 - 2}{3x_n^2}$$
$$= \frac{2x_n^3 + 2}{3x_n^2}.$$

Take $x_0 = 1.5$ as a first guess. Then

$$x_1 = \frac{2x_0^3 + 2}{3x_0^2} \doteq 1.2963$$
$$x_2 = \frac{2x_1^3 + 2}{3x_1^2} \doteq 1.26093$$
$$x_3 = \frac{2x_2^3 + 2}{3x_2^2} \doteq 1.25992$$
$$x_4 = \frac{2x_3^3 + 2}{3x_3^2} \doteq 1.25992.$$

The convergence is rapid and we get $x \doteq 1.25992$. □

Our main interest in Newton's method will be for calculating inverse function values. Suppose we are given a function f and that we wish to calculate $f^{-1}(a)$ for some number a in the domain of f^{-1}. We have

$$f(f^{-1}(a)) = a$$

so that
$$f(f^{-1}(a)) - a = 0.$$

Hence putting $x = f^{-1}(a)$ and solving $f(x) - a = 0$ for x gives us $f^{-1}(a)$.

EXAMPLE 10.7
Let $f(x) = x^3 + x + 1$, $x > 0$. Determine $f^{-1}(2)$. Notice that f^{-1} is well-defined since $f'(x) > 0$ for all x. According to the discussion in the text, we have to solve the equation $f(x) - 2 = 0$ for x, that is, we have to solve $x^3 + x + 1 - 2 = 0$. Newton's rule for this case is
$$x_{n+1} = \frac{2x_n^3 + 1}{3x_n^2 + 1}.$$

Take $x_0 = 1$ as a first guess. Then
$$x_1 = \frac{2x_0^3 + 1}{3x_0^2 + 1} = 0..725$$
$$x_2 = \frac{2x_1^3 + 1}{3x_1^2 + 1} \doteq 0.686047$$
$$x_3 = \frac{2x_2^3 + 1}{3x_2^2 + 1} \doteq 0.68234$$
$$x_4 = \frac{2x_3^3 + 1}{3x_3^2 + 1} \doteq 0.682328$$
$$x_5 = \frac{2x_4^3 + 1}{3x_4^2 + 1} \doteq 0.682328.$$

We conclude that $f^{-1}(2) \doteq 0.682328$.

10.2.2 The inverse of the logarithm function

We can now apply our knowledge of inverse functions to the logarithm function. Firstly, we note that
$$\frac{d}{dx}(\log x) = \frac{1}{x} > 0,$$

since the domain of $\log x$ is $(0, \infty)$. Hence $\log x$ is an increasing function on its domain, which implies that it is 1–1. This shows that the log function has an inverse. In order to conform with the notation of Theorem 10.2, let us put $f(x) = \log x$. Then the inverse of the logarithm function is f^{-1}. By Theorem 10.2 we have
$$(f^{-1})'(x) = \frac{1}{f'(f^{-1}(x))}. \tag{10.4}$$

Now $f'(x) = 1/x$, so the above equation can be written
$$(f^{-1})'(x) = \frac{1}{1/f^{-1}(x)}$$
$$= f^{-1}(x)$$

Hence the inverse of the log function is its own derivative. In other words

$$\frac{d}{dx}(\log^{-1} x) = \log^{-1} x.$$

Furthermore, $\log 1 = 0$, so that $\log^{-1} 0 = 1$. Hence $y(x) = \log^{-1}(x)$ satisfies the initial value problem

$$y'(x) = y(x), \quad y(0) = 1,$$

whose solution we know to be the exponential function. We would like to conclude that

$$\log^{-1}(x) = e^x. \tag{10.5}$$

This is in fact true, but it is important to realise that we are using a result from the theory of differential equations which states that initial value problems such as (10.5) have one and only one solution. The fact that exponentials and logarithms are inverses of each other proves the following theorem.

THEOREM 10.4
We have $e^{\log x} = x$ for all $x > 0$.

The relationship between the logarithmic and exponential functions is shown graphically in Figure 10.7.

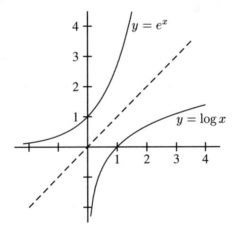

Figure 10.7: e^x as the inverse of $\log x$

10.2.3 Computing the exponential function

We have already computed e^x for various values of x using infinite series. We can now calculate it in a totally different way. Recall (page 207) that we can compute inverse functions by using Newton's method. Since the exponential function is the inverse of the logarithmic function, we can compute e^x by solving the equation

$$\log(e^x) - x = 0.$$

Let's try computing $y = \exp 1$, so that we have to solve $\log y - 1 = 0$ for y. Using Figure 10.7, we guess $y_0 = 2$. Then using

$$y_{n+1} = y_n - \frac{\log y_n - 1}{1/y_n},$$

we get

$$y_1 = y_0 - \frac{\log y_0 - 1}{1/y_0} \doteq 2.61371$$

$$y_2 = y_1 - \frac{\log y_1 - 1}{1/y_1} \doteq 2.71624$$

$$y_3 = y_2 - \frac{\log y_2 - 1}{1/y_2} \doteq 2.71828$$

$$y_4 = y_3 - \frac{\log y_3 - 1}{1/y_3} \doteq 2.71828.$$

Thus $e \doteq 2.71828$. If we are to be totally honest in the above calculation and not use built-in functions in calculators and/or computers, then the values $\log 2$, $\log 2.61371$, $\log 2.71624$ and $\log 2.71828$ will have to be calculated by a method such as Simpson's rule using the definition of the logarithm as an integral. Truly a heroic calculation!

The calculations of e by power series and by Simpson's rule and inverse functions are completely different, but produce exactly the same result. This is not an obvious fact. Rather, it is a consequence of the theory we have developed.

10.2.4 The air pressure problem

In Chapter 6 we solved the initial value problem

$$y'(x) = y(x), \quad y(0) = y_0$$

by using infinite series. Using the material developed in this chapter, we can now solve it by a quite different method. We have

$$\frac{y'(x)}{y(x)} = 1,$$

so that

$$\frac{d}{dx}(\log y) = 1.$$

Hence

$$\log y = x + c,$$

for some constant c. Thus

$$y = \log^{-1}(x + c)$$
$$= e^{x+c}$$
$$= Ae^x,$$

where $A = e^c$. Using the initial condition $y(0) = y_0$, we find $A = y_0$. Consequently, the solution to the initial value problem above is

$$y = y_0 e^x,$$

as we have previously shown.

EXAMPLE 10.8

When light passes vertically through a transparent substance, the rate at which the intensity $I(h)$ decreases is proportional to $I(h)$, where h represents the distance (in metres) travelled by the light. In clear seawater, it is found that the intensity of light 1 m below the surface is 25% of the intensity I_0 at the surface. What is the light intensity 5 m below the surface?

Solution. We can write
$$\frac{dI}{dh} = kI(h),$$
where k is a constant. This equation can be integrated to give
$$\log I(h) = kh + c,$$
$$I(h) = Ae^{kh},$$
where $A = e^c$. We have to find both A and k from the given information and then use these values to find $I(5)$. Firstly, we put $h = 0$, $I = I_0$ to get $A = I_0$, so that
$$I = I_0 e^{kh}.$$
Next, put $h = 1$, $I = 0.25 I_0$ to get
$$0.25 I_0 = I_0 e^k.$$
Hence $e^k = 0.25$. Finally, if $h = 5$, then
$$I = I_0 e^{5k}$$
$$= I_0 (e^k)^5$$
$$= I_0 \times (0.25)^5$$
$$\doteq 0.001 I_0,$$
or about 0.1% of the surface value.[1] □

We conclude this section by proving two results which make use of the inverse function relationships between exponentials and logarithms. First we prove a theorem on the convergence of the sequence $\{(1 + x/n)^n\}$.

THEOREM 10.5 *The sequence*
$$\left\{ \left(1 + \frac{x}{n}\right)^n \right\}$$
converges to e^x. In particular, the sequence
$$\left\{ \left(1 + \frac{1}{n}\right)^n \right\}$$
converges to e.

---

[1] This value may seem to run counter to what we expect from experience, but it only indicates that the human eye is able to function effectively over a vast range in light intensity. For example, the brightness of moonlight from a full moon is (on average) 0.0002% of the brightness of sunlight.

We have
$$\log\left(1+\frac{x}{n}\right)^n = n\log\left(1+\frac{x}{n}\right)$$
$$= n\left[\log\left(1+\frac{x}{n}\right) - \log 1\right]$$
$$= x\left[\frac{\log(1+x/n) - \log 1}{x/n}\right].$$

The thing to notice is that, as $n \to \infty$, the limit of the term
$$\frac{\log(1+x/n) - \log 1}{x/n}$$
is just
$$\log'(1) = 1.$$
Thus
$$\log\left(1+\frac{x}{n}\right)^n \to x$$
$$= \log e^x,$$
so that
$$\left(1+\frac{x}{n}\right)^n \to e^x.$$

The second result is the definition of irrational powers of a real number. Rational powers such as 2^3 and $2^{3/4}$ can be defined using ordinary arithmetic, but this is not the case for irrational powers such as $2^{\sqrt{3}}$. If r is any rational number, then by Theorem 10.4, we see that
$$a^r = e^{\log a^r} = e^{a\log r} \tag{10.6}$$

Equation (10.6) holds for all rational numbers r. We now define a^x for all x, rational or irrational, by
$$a^x = e^{a\log x}. \tag{10.7}$$

We call a^x the *general exponential function*.

EXAMPLE 10.9
Find $f'(x)$ if $f(x) = x^{x+1}$.
Solution. We write
$$\log f(x) = \log(x^{x+1})$$
$$= (x+1)\log x.$$

DIfferentiating both sides gives
$$\frac{f'(x)}{f(x)} = \frac{x+1}{x} + \log x$$
$$= 1 + \frac{1}{x} + \log x.$$

Consequently,

$$f'(x) = f(x)\left(1 + \frac{1}{x} + \log x\right)$$
$$= x^{x+1}\left(1 + \frac{1}{x} + \log x\right).$$

EXERCISES 10.2

Simplify the following:

1. $e^{\log x}$
2. $\log(e^x)$
3. $e^{2\log x}$
4. $e^{-\log x}$
5. $e^{x - 2\log x}$
6. $\log(xe^x)$

Differentiate the following functions:

7. $f(x) = x^x$
8. $f(x) = x^{\sin x}$
9. $f(x) = (\cos x)^x$
10. $f(x) = 2^{\sqrt{x}}$
11. $f(x) = (\log x)^{\sqrt{x}}$

Apply Newton's method in Exercises 12–17 to find a root of the given equation in the indicated interval.

12. $x^2 - 2 = 0$, $[1, 2]$
13. $x^3 - 2 = 0$, $[1, 2]$
14. $x - \cos x = 0$, $[0, 2]$
15. $x^3 - 4x + 1$, $[1, 3]$
16. $\sin x - x^2$, $[0, 2]$
17. $x^2 - 2x - 5 = 0$, $[1, 3]$

18. Use Newton's method to find a value of x for which

$$\int_0^x \frac{t^2}{1 + t^2}\, dt = \frac{1}{2}.$$

In Exercises 19–22, apply Newton's method to find the value of $f^{-1}(a)$ at the given point a.

19. $f(x) = x^3$, $a = 2$
20. $f(x) = e^{2x}$, $a = 2$
21. $f(x) = \ln x$, $a = 3$
22. $f(x) = \sin x$, $x \in [0, 1]$, $a = 0.5$

23. The ancient Babylonians discovered a way of extracting square roots about 4000 years ago. In modern notation, their method of finding $\sqrt{a}$ was to make an initial guess x for $\sqrt{a}$ and then to take the average of x and a/x as a better guess for $\sqrt{a}$. This process was repeated until the desired accuracy was reached. Use this method to evaluate $\sqrt{6}$ to 4 decimal places.

24. Show that the Babylonian method in the above example is equivalent to using Newton's method on the equation $x^2 - a = 0$.

25. A bacteria culture increases by a factor of 6 in 10 hours. How long did it take the number of bacteria to double?

26. In 1987, the world's population was 5 billion. At that time the population was increasing at a rate of 380 000 people per day. When would the world's population be expected to double? To triple?

27. Radioactive elements decay at a rate proportional to the amount of the element present. Suppose a sample of a radioactive element Ra^{222} decays to 58% of its original amount after 3 days.

    (a) How long does it take for the sample to decay to half of its original value? This is called the *half-life* of the element.

    (b) How long does it take for the sample to decay to 20% of its original value?

Atmospheric air contains carbon, almost all of it in the form of carbon dioxide. Most of the atmospheric carbon occurs as C^{12}, but there is a small percentage of radioactive carbon (C^{14}). This radioactive carbon decays with a half life of 5600 years. However, the ratio of C^{14} to C^{12} in the atmosphere remains approximately constant, because the C^{14} is continually replenished by the action of cosmic rays on nitrogen in the upper atmosphere. Living organisms absorb both the C^{12} and the C^{14} and the ratios of these substances in a living organism is the same as in the atmosphere. Once the organism dies, the C^{14} decays and is not replaced, so that the ratio of C^{14} to C^{12} changes with time and this ratio can give an indication of the time since death. This is the basis of *carbon dating*.

28. A bone excavated from an archeological site is found to contain 1% of the original amount of C^{14}. Determine the age of the bone.

29. Assume that the temperature of a cup of coffee obeys Newton's law of cooling. Suppose the coffee has a temperature of 95°C when freshly poured and has cooled to 90°C one minute later. If the room temperature is 25°C, determine when the coffee will have cooled to 50°C.

30. A body at a temperature of 32°C is discovered at midnight in a room of temperature 25°C. The body is speedily taken to the morgue where the temperature is kept at 4°C. After one hour, the temperature of the body has fallen to 25°C. Estimate the time of death.

10.3 THE OSCILLATION PROBLEM AGAIN

In Chapter 7 we solved the differential equation

$$y'' + y = 0, \quad y(0) = 0, \quad y'(0) = 1 \tag{10.8}$$

by means of infinite series. We will now solve it using some of the ideas in this and the previous chapter. We begin by expressing the left hand side of equation (10.8) as a first order derivative. The trick in doing this is to multiply the equation by y'. This gives

$$y''y' + yy' = 0.$$

Using the chain rule, we find

$$\frac{d}{dt}\left(\tfrac{1}{2}(y')^2\right) = \tfrac{1}{2} \times 2y'y'' = y''y'$$

and that

$$\frac{d}{dt}\left(\tfrac{1}{2}y^2\right) = \tfrac{1}{2} \times 2yy' = yy',$$

then the above equation can be written

$$\frac{d}{dt}\left(\tfrac{1}{2}(y')^2 + \tfrac{1}{2}y^2\right) = 0$$

and so
$$\tfrac{1}{2}(y')^2 + \tfrac{1}{2}y^2 = c, \tag{10.9}$$
where c is a constant.

This last equation has a physical interpretation, which is disguised by the way it has been written. If we take the differential equation
$$my'' + ky = 0$$
for the particle of mass m on a spring with spring constant k, the above condition becomes
$$\tfrac{1}{2}m(y')^2 + \tfrac{1}{2}ky^2 = c_1. \tag{10.10}$$
This says that the quantity on the left hand side is constant throughout the motion. The constant is called the *energy* and so we can say that the energy is conserved in the motion. The first term is the *kinetic energy* of the particle while the second is its *potential energy*. The left hand side of equation (10.10) is the total energy of the particle. During motion, energy is changing from one form to the other, but the total energy is constant. This statement assumes that there is no friction in the motion.

By using the initial conditions $y(0) = 0$ and $y'(0) = 1$ in equation (10.9), we find $c = \tfrac{1}{2}$ and so
$$(y')^2 + y^2 = 1.$$
This gives us two possibilities:
$$y' = \pm\sqrt{1 - y^2}.$$
To decide which of these to take, we use the initial conditions again. Since $y'(0) > 0$, we discard the negative sign. Hence
$$y' = \sqrt{1 - y^2},$$
which we write as
$$\frac{1}{\sqrt{1 - y^2}} \frac{dy}{dt} = 1. \tag{10.11}$$
Following our method in the last chapter, we define a function F whose derivative satisfies
$$F'(y) = \frac{1}{\sqrt{1 - y^2}}.$$
Then equation (10.11) becomes
$$\frac{d}{dt}\Big(F(y(t))\Big) = 1$$
and hence $F(y(t)) = t + a$, for some constant a. The initial condition $y(0) = 0$ allows us to find $a = F(0)$, so the solution is
$$F(y(t)) = t + F(0).$$
All that remains is to find F. We might expect to find an algebraic function for F, but this turns out not to be possible, so we to use the device we developed in the previous chapter and define F in terms of an integral. We define
$$F(x) = \int_{x_0}^{x} \frac{1}{\sqrt{1 - u^2}} \, du.$$

Here x_0 can be any constant. A judicious choice simplifies things. If $x_0 = 0$, then $F(0) = 0$ and the above solution becomes
$$F(y(t)) = t, \qquad (10.12)$$
with
$$F(x) = \int_0^x \frac{1}{\sqrt{1-u^2}}\, du, \quad -1 \leqslant x \leqslant 1.$$

The restriction $-1 \leqslant x \leqslant 1$ is necessary because the integrand (that is, the function being integrated) has vertical asymptotes at $u = \pm 1$.

We already know from Chapter 7 that the solution $y(t)$ of the initial value problem (10.8) is $y(t) = \sin t$, so that we can write equation (10.12) as
$$F(\sin t) = t. \qquad (10.13)$$

A function F which has an inverse F^{-1} satisfies the equation $F(F^{-1}(t)) = t$ for all t in the range of F, and so equation (10.13) suggests that we write $F^{-1}(t) = \sin t$, or equivalently, that $F(t) = \sin^{-1}(t)$. However, we should not be too hasty in concluding that $F^{-1}(t) = \sin t$. A function and its inverse are both 1–1 functions and certainly $\sin t$ is *not* a 1–1 function. What is true though, is that a suitable restriction of $\sin t$ will be 1–1 and it is this restricted function that we will have to identify with $F^{-1}(t)$. Since $F(0) = 0$, we must have $F^{-1}(0) = 0$ and so it is the restriction of $\sin t$ to the interval $[-\pi/2, \pi/2]$ which has to be identified with $F^{-1}(t)$.

We can also show directly from the definition of F and the use of Theorem 10.2 that F^{-1} satisfies the initial value problem (10.8). As an exercise, you should show that F is 1–1. Consequently, F has an inverse F^{-1}. For ease of notation let $F^{-1} = S$. The derivative of F is given by
$$F'(t) = \frac{1}{\sqrt{1-t^2}}.$$

From Theorem 10.2, the derivative of the inverse is
$$S'(t) = \frac{1}{F'(S(t))}$$
$$= \frac{1}{1/\sqrt{1-[S(t)]^2}}$$
$$= \sqrt{1-[S(t)]^2}.$$

Then, by the chain rule
$$S''(t) = -\frac{S(t)S'(t)}{\sqrt{1-[S(t)]^2}}$$
$$= -S(t).$$

Also, $F(0) = 0$, so $S(0) = 0$ and from the above equation for $S'(t)$,
$$S'(0) = \sqrt{1-[S(0)]^2} = 1.$$

Putting all this together, we see that S satifies the initial value problem
$$S'' + S = 0, \quad S(0) = 0, \quad S'(0) = 1.$$

Hence, as before,
$$F^{-1}(t) = S(t) = \sin t, \quad -\frac{\pi}{2} \leqslant t \leqslant \frac{\pi}{2}.$$

The function $F(t)$ is the inverse of $S(t)$ and is called the *inverse sine function*. The function F has domain $[-1, 1]$ and range $[-\pi/2, \pi/2]$ and is denoted by $\sin^{-1} x$. This terminology is not strictly correct: $F(t)$ is *not* the inverse of $\sin t$, it is the inverse of the restriction of $\sin t$ to the domain $[-\pi/2, \pi/2]$. However, this notation is well established and the alternative is too clumsy in use.

What more can we say about the properties of the function F? We begin by plotting its graph using *Mathematica*. The basic command is

```
Plot[Evaluate[Integrate[1/Sqrt[1-t^2],{t,0,x}]],{x,-1,1}]
```

We have restricted x to the domain $[-1, 1]$ of F. To get a nicer looking plot, try the following instruction.

```
Plot[Evaluate[Integrate[1/Sqrt[1-t^2],{t,0,x}]],{x,-1,1},
   PlotStyle->RGBColor[1,0,0],AspectRatio->Automatic,
   Ticks->{{-1,1},{-1.5,-1,-.5,.5,1,1.5}}]
```

The graph is shown in Figure 10.8.

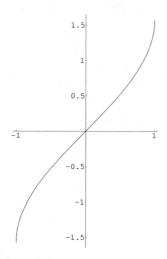

Figure 10.8: A *Mathematica* plot of the graph of F

The plot suggests that $F(1) \approx 1.6$ and $F(-1) \approx -1.6$. We have shown that the exact values are $F(1) = \pi/2$ and $F(-1) = -\pi/2$. Clearly $F(0) = 0$ from the way we have defined F. Also, by the Fundamental Theorem of the Calculus,

$$F'(t) = \frac{1}{\sqrt{1-t^2}}$$

is always positive and so F is an increasing function. The second derivative is

$$F''(t) = \frac{t}{(1-t^2)^{3/2}}$$

and hence there is a point of inflexion at $t = 0$.

218 INVERSE FUNCTIONS

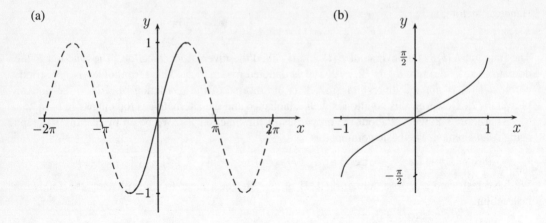

Figure 10.9: (a) A restricted domain for sin x. (b) The inverse sine function.

10.4 INVERSE TRIGONOMETRIC FUNCTIONS

In the previous section we met the inverse sine function. In this section we will discuss some of the properties of this and other inverse trigonometric functions. Figure 10.9(a) shows the way in which we can restrict the function $\sin x$ to a part of its domain on which it is 1–1. The inverse of this restricted function is known as the *inverse sine function*. The graph of $\sin^{-1} x$ is shown in Figure 10.9(b). The definition of the inverse sine function is given in the box below.

DEFINITION 10.3 The inverse sine function
The inverse of the function
$$f(x) = \sin x, \quad -\frac{\pi}{2} \leqslant x \leqslant \frac{\pi}{2}$$
is called the **inverse sine function**. *We denote $f^{-1}(x)$ by $\sin^{-1} x$. The inverse function relations are*
$$\sin(\sin^{-1} x) = x, \quad -1 \leqslant x \leqslant 1$$
$$\sin^{-1}(\sin x) = x, \quad -\frac{\pi}{2} \leqslant x \leqslant \frac{\pi}{2}$$

We can also define the *inverse cosine function*. Once again, we have to restrict the function to an interval on which it is 1–1. The conventional choice is $[0, \pi]$.

DEFINITION 10.4 The inverse cosine function
The inverse of the function
$$f(x) = \cos x, \quad 0 \leqslant x \leqslant \pi$$
is called the **inverse cosine function**. *We denote $f^{-1}(x)$ by $\cos^{-1} x$. The inverse function relations are*
$$\cos(\cos^{-1} x) = x, \quad -1 \leqslant x \leqslant 1$$
$$\cos^{-1}(\cos x) = x, \quad 0 \leqslant x \leqslant \pi$$

Figure 10.10(a) shows the way in which we restrict the function $\cos x$ to a part of its domain on which it is 1–1. The graph of $\cos^{-1} x$ is shown in Figure 10.10(b).

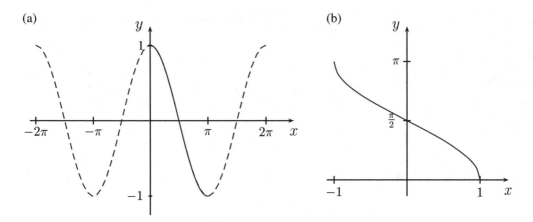

Figure 10.10: (a) A restricted domain for $\cos x$. (b) The inverse cosine function.

10.4.1 The derivatives of the inverse trigonometric functions

The derivatives of the inverse trigonometric functions can be found using Theorem 10.2. For the inverse sine function we have

$$(\sin^{-1})'(x) = \frac{1}{\sin'(\sin^{-1} x)}$$
$$= \frac{1}{\cos(\sin^{-1} x)}$$
$$= \frac{1}{\sqrt{1 - \sin^2(\sin^{-1} x)}}$$
$$= \frac{1}{\sqrt{1 - x^2}}.$$

To find the derivative of $\cos^{-1} x$, there are two ways to proceed. Firstly, let $y = \cos^{-1} x$. Then $x = \cos y$ and

$$\frac{d}{dy}(\cos y) = -\sin y. \tag{10.14}$$

We can write $\sin y$ in terms of $\cos y$ by using the identity $\cos^2 y + \sin^2 y = 1$. We have

$$\sin^2 y = 1 - \cos^2 y,$$

so that

$$\sin y = \pm\sqrt{1 - \cos^2 y}.$$

To decide which sign to take, we have to use the fact that the slope of the graph of $\cos^{-1} x$ in Figure 10.10(b) is negative, so that the left hand side of equation (10.14) must be negative and hence

220 INVERSE FUNCTIONS

$\sin y$ must be positive. We take $\sin y = \sqrt{1 - \cos^2 y}$. Then

$$\frac{d}{dy}(\cos y) = -\sqrt{1 - \cos^2 y}$$
$$= -\sqrt{1 - x^2}.$$

Since
$$\frac{dy}{dx} = 1 \Big/ \frac{dx}{dy},$$

we have
$$\frac{d}{dx}(\cos^{-1} x) = -\frac{1}{\sqrt{1 - x^2}}.$$

Alternatively, we can use Theorem 10.2 directly:

$$(\cos^{-1})'(x) = \frac{1}{\cos'(\cos^{-1} x)}$$
$$= \frac{1}{-\sin(\cos^{-1} x)}$$
$$= -\frac{1}{\sqrt{1 - \cos^2(\cos^{-1} x)}}$$
$$= -\frac{1}{\sqrt{1 - x^2}}.$$

EXAMPLE 10.10
By differentiating
$$\sin^{-1} x + \cos^{-1} x,$$

show that $\sin^{-1} x + \cos^{-1} x$ is a constant. Hence deduce that

$$\sin^{-1} x + \cos^{-1} x = \frac{\pi}{2}.$$

Solution. We have

$$\frac{d}{dx}(\sin^{-1} x + \cos^{-1} x) = \frac{1}{\sqrt{1 - x^2}} + \left(-\frac{1}{\sqrt{1 - x^2}}\right)$$
$$= 0.$$

Hence
$$\sin^{-1} x + \cos^{-1} x = c,$$

where c is a constant. Put $x = 0$. Then

$$c = \sin^{-1} 0 + \cos^{-1} 0 = \frac{\pi}{2},$$

so that
$$\sin^{-1} x + \cos^{-1} x = \frac{\pi}{2}.$$

This result is the reason for the co-terminology for trigonometric functions. The inverse of a function and its co-function sum to $\pi/2$. Thus each has the same derivative apart from a sign.

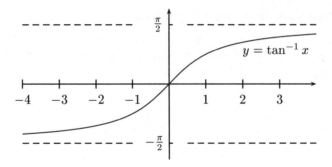

Figure 10.11: The graph of $\tan^{-1} x$

10.5 OTHER INVERSE TRIGONOMETRIC FUNCTIONS

In Chapter 7, we defined four other trigonometric functions—$\tan x$, $\sec x$, $\csc x$ and $\cot x$. Each of these, suitably restricted, has an inverse function. The only one we consider here is the inverse tangent. We leave the others to be done as exercises. Since $\tan x = \sin x / \cos x$, the graph will have vertical asymptotes at the points where $\cos x = 0$, that is, at multiples of $\pi/2$. Since the function is not 1–1, we need to restrict the domain in order to construct an inverse function. The natural choice is to take $-\pi/2 < x < \pi/2$. The range of the restricted function is $\mathbb{R}$, so there is an inverse (denoted by $\tan^{-1} x$) with domain $\mathbb{R}$ and range $-\pi/2 < x < \pi/2$ (Figure 10.11). The formal definition is the following.

> **DEFINITION 10.5 The inverse tangent**
> *The inverse of the function*
> $$f(x) = \tan x, \quad -\frac{\pi}{2} < x < \frac{\pi}{2}$$
> *is called the* **inverse tangent function**. *We denote* $f^{-1}(x)$ *by* $\tan^{-1} x$. *The inverse function relations are*
> $$\tan(\tan^{-1} x) = x, \quad x \in \mathbb{R}$$
> $$\tan^{-1}(\tan x) = x, \quad -\frac{\pi}{2} < x < \frac{\pi}{2}.$$

To compute the derivative of $\tan^{-1} x$ we use Theorem 10.2:

$$\begin{aligned}
(\tan^{-1})'(x) &= \frac{1}{\tan'(\tan^{-1} x)} \\
&= \frac{1}{\sec^2(\tan^{-1} x)} \\
&= \frac{1}{1 + \tan^2(\tan^{-1} x)} \\
&= \frac{1}{1 + x^2}.
\end{aligned}$$

10.5.1 Computing inverse trigonometric functions

Apart from a few special angles, we have to resort to numerical methods, such as Newton's method to compute values of the inverse trigonometric functions. To compute a value of $\sin^{-1} x$, for instance, we have to solve the equation $\sin y = x$ for the value y.

EXAMPLE 10.11
Compute $\tan^{-1} 2$ to 4 decimal places.
Solution. Put $x = \tan^{-1} 2$. Then $\tan x = 2$. We put $f(x) = \tan x - 2$ and use Newton's method in the form
$$x_{n+1} = x_n - \frac{f(x_n)}{f'(x_n)}.$$
We know that $\tan(\pi/4) = 1$, so take as a starting value something a bit bigger than $\pi/4$, say $x_0 = 1$. Then
$$x_1 = x_0 - \frac{\tan x_0 - 2}{\sec^2 x_0} \doteq 1.1292$$
$$x_2 = x_1 - \frac{\tan x_1 - 2}{\sec^2 x_1} \doteq 1.10813$$
$$x_3 = x_2 - \frac{\tan x_2 - 2}{\sec^2 x_2} \doteq 1.10715$$
$$x_4 = x_3 - \frac{\tan x_3 - 2}{\sec^2 x_3} \doteq 1.10715$$

We conclude $\tan^{-1} 2 \doteq 1.1071$ One minor point: How do we calculate the values of $\tan x$ and $\sec x$? In principle, we have to use infinite series. Of course, nobody would routinely calculate these functions in such a laborious fashion, since they are built into any reasonable calculating device. In fact, so are the values of $\tan^{-1} x$. However, the method described above is essentially that used by the calculator, which does not have some mysterious method of its own for doing the calculation. The point of this discussion is not to give practical methods for calculating function values, but rather to lay bare the processes which have to be taken into account when designing algorithms for calculators and computers. These functions are built into the calculating device precisely because they arise in a large number of applications.

10.5.2 CONCLUSIONS

The significance of the inverse trigonometric functions really lies in their derivatives. The derivative of a trigonometric function is another trigonometric function, but this is not true of the inverse trigonometric functions. Our discussion in this chapter yields two results that we could not have obtained in other ways. These are

$$\int \frac{1}{\sqrt{1-x^2}}\, dx = \sin^{-1} x + c \tag{10.15}$$

and

$$\int \frac{1}{1+x^2}\, dx = \tan^{-1} x + c. \tag{10.16}$$

These integrals arise in many problems. As we have already seen, the first arises in oscillation problems when we use energy methods to find the solution.

EXERCISES 10.5

Determine the exact value of the following:

1. $\cos^{-1} 0$
2. $\sin^{-1} 1$
3. $\cos^{-1} \frac{1}{2}$
4. $\tan^{-1} \sqrt{3}$
5. $\sin^{-1}(\sin \frac{7}{4}\pi)$
6. $\sin(\cos^{-1} \frac{1}{2})$
7. $\cos\left(\cos^{-1}\left(\frac{1}{2}\sqrt{2}\right)\right)$
8. $\tan^{-1}(\cos 0)$
9. $\cos^{-1}(\tan 0)$
10. $\tan^{-1}(-1)$
11. $\sin^{-1}(\cos \frac{3}{2}\pi)$
12. $\sin(\tan^{-1} 1)$

If $x > 0$, calculate the following:

13. $\sin(\tan^{-1} x)$
14. $\cos(\tan^{-1} x)$
15. $\tan(\tan^{-1} x)$

If $0 < x < 1$, calculate the following:

16. $\sin(\cos^{-1} x)$
17. $\cos(\cos^{-1} x)$
18. $\tan(\cos^{-1} x)$
19. $\sin(\sin^{-1} x)$
20. $\cos(\cos^{-1} x)$
21. $\tan(\tan^{-1} x)$

Differentiate the following functions:

22. $f(x) = \cos^{-1}(x^2 - 1)$
23. $f(x) = \tan^{-1}\left(\frac{1}{x}\right)$
24. $f(x) = e^x \cos^{-1} x$
25. $y = \tan^{-1}(\sin x)$
26. $y = \dfrac{x}{\sqrt{4-x^2}} - \sin^{-1}\left(\dfrac{x}{2}\right)$
27. $y = \sqrt{9-x^2} + 3\sin^{-1}\left(\dfrac{x}{3}\right)$
28. $f(r) = \sin^{-1}\sqrt{1-r^2}$
29. $f(x) = \tan^{-1}(x^{2x})$

30. Differentiate
$$f(x) = \sin^{-1}\left(\frac{x+b}{a}\right).$$
Hence evaluate the integral
$$\int \frac{dx}{\sqrt{a^2 - (x+b)^2}}.$$

31. Differentiate
$$f(x) = \frac{1}{a}\tan^{-1}\left(\frac{x+b}{a}\right).$$
Hence evaluate the integral
$$\int \frac{dx}{a^2 + (x+b)^2}.$$

CHAPTER 11
HYPERBOLIC FUNCTIONS

Much of this book has been about the need to introduce functions in order to solve particular problems. The collection of functions we have had to introduce includes algebraic formulas, roots, exponential and logarithmic functions, trigonometric functions and inverse trigonometric functions. These functions, together with all combinations of them, are called *elementary functions*. The word *elementary* is used here in its sense of *fundamental* or *basic*, not in the sense of *easy*. Indeed, as we have seen, there is nothing particularly easy about most of these functions. These functions provide solutions to problems such as projectile motion, growth and decay and oscillation problems, but there are problems for which these functions are insufficient to provide a solution. The circular motion in the *Tower of Terror* provides such an example and we shall consider it in Chapter 13. As well as the elementary functions we have already considered, there is another group of elementary functions which play a sufficiently important role in many problems to have been given specific names. These are the *hyperbolic functions* and their inverses.

11.1 HYPERBOLIC FUNCTIONS

The hyperbolic functions are defined as certain combinations of exponential functions. If you think about the links between complex exponential functions and the trigonometric functions, you will not be surprised to discover that the hyperbolic functions have many properties which are similar to trigonometric functions.

To motivate the definitions let us consider the equation $y'' + y = 0$, which is an example of undamped simple harmonic motion. Its auxiliary equation is $m^2 + 1 = 0$ with roots $m = \pm i$. In real form, the solution is

$$y = C \cos t + D \sin t, \tag{11.1}$$

which we can write in complex form as as

$$y = C \left(\frac{e^{it} + e^{-it}}{2} \right) + D \left(\frac{e^{it} - e^{-it}}{2i} \right). \tag{11.2}$$

Now let's keep this example in mind and consider the case

$$y'' - y = 0.$$

The auxiliary equation is $m^2 - 1 = 0$ with roots $m = \pm 1$, so the solution is

$$y = Ae^t + Be^{-t}.$$

We can regroup the exponential terms in a similar way to that of equation (11.2) by rewriting it as follows:

$$\begin{aligned} y &= \left(\frac{A}{2} + \frac{A}{2}\right) e^t + \left(\frac{B}{2} + \frac{B}{2}\right) e^{-t} \\ &= \left(\frac{A}{2} + \frac{A}{2}\right) e^t + \left(\frac{B}{2} + \frac{B}{2}\right) e^{-t} + \left(\frac{A}{2} - \frac{A}{2}\right) e^{-t} + \left(\frac{B}{2} - \frac{B}{2}\right) e^t \\ &= (A + B) \left(\frac{e^t + e^{-t}}{2}\right) + (A - B) \left(\frac{e^t - e^{-t}}{2}\right) \\ &= C \left(\frac{e^t + e^{-t}}{2}\right) + D \left(\frac{e^t - e^{-t}}{2}\right) \end{aligned} \qquad (11.3)$$

where $C = A + B$ and $D = A - B$. There is a marked similarity between the form of this last result and equation (11.2). We give names to the two combinations of exponential functions occurring in equation (11.3) by defining

$$\cosh t = \frac{e^t + e^{-t}}{2}$$

$$\sinh t = \frac{e^t - e^{-t}}{2}.$$

Then the solution (11.3) becomes

$$y = C \cosh t + D \sinh t. \qquad (11.4)$$

The function $\cosh t$ is called the *hyperbolic cosine* of t, while $\sinh t$ is called the *hyperbolic sine* of t. They are often pronounced as *cosh* and *shine* (or *sinch* by Americans). The similarity between the form of equation (11.4) and that of equation (11.1) is obvious, but this is simply because of the notation. However, this notation has been chosen for a reason. It turns out that there are many similarities between the trigonometric functions and the hyperbolic functions, some of which we shall explore below.

EXAMPLE 11.1
We consider a case of strongly damped motion given by the equation

$$y'' + 6y' + 4y = 0.$$

The auxiliary equation is $m^2 + 6m + 4 = 0$ with roots $m = -3 \pm \sqrt{5}$, so the solution is

$$y = e^{-3t}(A e^{\sqrt{5}t} + B e^{-\sqrt{5}t}).$$

We can express this in terms of hyperbolic functions by writing

$$\begin{aligned} y &= e^{-3t} \left[\frac{A+B}{2}\left(e^{\sqrt{5}t} + e^{-\sqrt{5}t}\right) + \frac{A-B}{2}\left(e^{\sqrt{5}t} - e^{-\sqrt{5}t}\right)\right] \\ &= e^{-3t} \left[C\left(\frac{e^{\sqrt{5}t} + e^{-\sqrt{5}t}}{2}\right) + D\left(\frac{e^{\sqrt{5}t} - e^{-\sqrt{5}t}}{2}\right)\right], \\ &= e^{-3t}(C \cosh \sqrt{5}t + D \sinh \sqrt{5}t). \end{aligned}$$

where $C = A + B$ and $D = A - B$. □

In general, if the auxiliary equation of the differential equation
$$ay'' + by' + cy = 0$$
has real roots $p \pm q$, then the solution can be expressed in the form
$$y(t) = e^{pt}(A \cosh qt + B \sinh qt). \tag{11.5}$$

EXAMPLE 11.2
Solve the equation
$$y'' - 4y = 0.$$
The auxiliary equation is $m^2 - 4 = 0$; it has roots ± 2. The general solution is therefore
$$y = A \cosh 2t + B \sinh 2t. \qquad \square$$

By analogy with the trigonometric functions, we define four other hyperbolic functions. These are
$$\tanh x = \frac{\sinh x}{\cosh x}, \quad \coth x = \frac{1}{\tanh x}$$
$$\operatorname{sech} x = \frac{1}{\cosh x}, \quad \operatorname{csch} x = \frac{1}{\sinh x}.$$

Since the hyperbolic functions are combinations of exponential functions, we could still do all of the problems in this book without introducing them. However, they are a very useful convenience in many circumstances and we shall use them as the need arises. There is a simple occcurrence of the hyperbolic cosine that we see almost every day. If a heavy cable such as a power line is strung between two points, the curve it hangs in will be a hyperbolic cosine.

EXERCISES 11.1

Find general solutions of the differential equations in exercises 1–4 in terms of the hyperbolic sine and cosine functions.

1. $y'' - 2y = 0$
2. $y'' - 9y = 0$
3. $y'' + 6y' + 5y = 0$
4. $x'' - 2x' - 2x = 0$

5. Find general solutions of the differential equations in exercises 1–4 in terms of exponential functions. Verify that the two sets of answers you have are equivalent.

11.2 PROPERTIES OF THE HYPERBOLIC FUNCTIONS

It follows from the definitions that $\cosh 0 = 1$ and $\sinh 0 = 0$. These are the only values of these functions which can be computed exactly. The graphs can be plotted from a knowledge of the graphs of e^x and e^{-x}. They are shown in Figure 11.1.

The derivatives of these functions show a similarity to those of the trigonometric functions. For the hyperbolic cosine, we have
$$\frac{d}{dx}(\cosh x) = \frac{1}{2}\frac{d}{dx}(e^x + e^{-x})$$
$$= \frac{1}{2}(e^x - e^{-x})$$
$$= \sinh x.$$

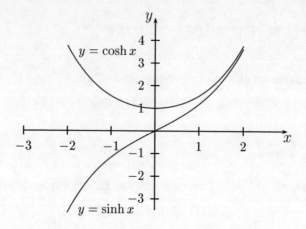

Figure 11.1: Graphs of $y = \cosh x$ and $y = \sinh x$

Notice that, unlike the corresponding trigonometric result, there is no minus sign. The result for the derivatives of the hyperbolic sine and hyperbolic tangent are obtained similarly. We find

$$\frac{d}{dx}(\sinh x) = \cosh x,$$
$$\frac{d}{dx}(\tanh x) = \operatorname{sech}^2 x.$$

The hyperbolic functions satisfy many identities similar to those of the trigonometric functions.

EXAMPLE 11.3
Show that
$$\cosh^2 x - \sinh^2 x = 1.$$

Solution. We just use the definition:
$$\cosh^2 x - \sinh^2 x = \left[\tfrac{1}{2}(e^x + e^{-x})\right]^2 - \left[\tfrac{1}{2}(e^x - e^{-x})\right]^2$$
$$= 1. \qquad \square$$

There is a rule for going from a trigonometric identity to a hyperbolic identity, as long as differentiation and integration are not involved. Suppose we have a given trigonometric identity. The rule for getting the corresponding hyperbolic identity is this:

- Change all trigonometric functions to the corresponding hyperbolic functions—sin to sinh, cos to cosh, etc.

- If the trigonometric identity contains the product of two sines, change the sign in the corresponding hyperbolic identity.

- Be aware that the product of two sines may be disguised. For instance if a term such as $\tan^2 \theta$ appears in a trigonometric identity, then since

$$\tan^2 \theta = \frac{\sin^2 \theta}{\cos^2 \theta},$$

we will have to change the sign in the corresponding hyperbolic identity.

EXAMPLE 11.4
We have
$$\cos(A+B) = \cos A \cos B - \sin A \sin B,$$
so the hyperbolic identity is
$$\cosh(A+B) = \cosh A \cosh B + \sinh A \sinh B. \qquad \square$$

This rule is a consequence of the fact that both the trigonometric functions and the hyperbolic functions can be expressed in exponential form. We have
$$\cos x = \frac{e^{ix} - e^{-ix}}{2} = \cosh ix,$$
and
$$\sin x = \frac{e^{ix} - e^{-ix}}{2i} = -i \sinh ix.$$

So, for example, since
$$\cos^2 x + \sin^2 x = 1,$$
we have
$$\cosh^2 ix + (-i \sinh ix)^2 = 1,$$
$$\cosh^2 ix - \sinh^2 ix = 1.$$

Hence
$$\cosh^2 y - \sinh^2 y = 1,$$
where $y = ix$.

We have here another example of results about real functions which are explained by the use of complex numbers. Without complex numbers, the similarity between trigonometric and hyperbolic functions would have been inexplicable.

EXERCISES 11.2

Prove the following identities.

1. $\cosh 0 = 1$
2. $\sinh 0 = 0$
3. $\cosh(-x) = \cosh x$
4. $\sinh(-x) = -\sinh x$
5. $\tanh(-x) = -\tanh$
6. $\operatorname{sech}^2 x = 1 - \tanh^2 x$
7. $\cosh(x - y) = \cosh x \cosh y - \sinh x \sinh y$
8. $\sinh(x + y) = \sinh x \cosh y + \cosh x \sinh y$
9. $\sinh(x - y) = \sinh x \cosh y - \cosh x \sinh y$
10. $\sinh 2x = 2 \sinh x \cosh x$
11. $\cosh 2x = 2 \sinh^2 x + 1$

12. Show that $e^x = \cosh x + \sinh x$ and $e^{-x} = \cosh x - \sinh x$. Hence show that
$$(\cosh x + \sinh x)^n = \cosh nx + \sinh nx$$
and
$$(\cosh x - \sinh x)^n = \cosh nx - \sinh nx$$
for all $n, x \in \mathbb{R}$.

Find the derivative of the following functions.

13. $f(x) = \cosh(2x + 1)$
14. $f(x) = \sinh x^3$
15. $f(x) = \dfrac{1}{2} \log |\tanh x|$
16. $f(x) = \ln \sinh 2x$
17. $f(x) = \cosh^3 x$
18. $f(x) = x - \tanh x$
19. $f(x) = \cosh x^3$
20. $f(x) = \cosh \log x$
21. $f(x) = \log \operatorname{sech} x$
22. $f(x) = \sinh \left(\dfrac{1}{x}\right)$
23. $f(x) = \sinh(\sin x)$
24. $f(x) = \dfrac{\sinh x}{\cosh x - 1}$
25. $f(x) = \ln |\sinh ax|$
26. $f(x) = 2 \tanh \dfrac{x}{2}$

27. A body of mass m falls from rest under the action of gravity. It encounters an air resistance proportional to the square of its velocity. Satisfy yourself that the motion is described by the initial value problem
$$m \frac{dv}{dt} = mg - kv^2, \quad v(0) = 0,$$
where k is a constant which depends on the properties of the particular body.

 (a) Show that
$$v = \sqrt{\frac{mg}{k}} \tanh \left(\sqrt{\frac{gk}{m}}\, t\right)$$
 is a solution of this initial value problem.

 (b) What value does the velocity approach as t gets large? What physical phenomenon is occurring here?

 (c) Let $s(t)$ be the distance of the body from some fixed point at time t. If it is known that $s(0) = s_0$, find an explicit expression for s.

11.3 INVERSE HYPERBOLIC FUNCTIONS

The hyperbolic functions have inverses which are useful for computing integrals. The function $\cosh x$ is not 1–1, but if we restrict the domain to $[0, \infty)$, we do get a 1–1 function. The inverse of this restricted function is denoted by $\cosh^{-1} x$ and is called (somewhat incorrectly) the *inverse hyperbolic cosine function*. It has domain $[1, \infty)$ and range $[0, \infty)$ (Figure 11.2).

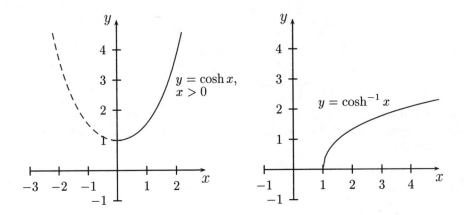

Figure 11.2: Graphs of $y = \cosh x$, $x > 0$ and $y = \cosh^{-1} x$

To find the derivative, let $f(x) = \cosh x$ for $x > 0$. Then

$$f'(x) = \sinh x$$
$$= \sqrt{\cosh^2 x - 1},$$

since $\sinh x > 0$.

The domain of the inverse function is $[1, \infty)$, so for $x \geqslant 1$ we have

$$\frac{d}{dx}(\cosh^{-1} x) = (f^{-1})'(x)$$
$$= \frac{1}{f'(f^{-1}x)}$$
$$= \frac{1}{\sinh(f^{-1}x)}$$
$$= \frac{1}{\sqrt{\cosh^2(f^{-1}x) - 1}}$$
$$= \frac{1}{\sqrt{\cosh^2(\cosh^{-1} x) - 1}}$$
$$= \frac{1}{\sqrt{x^2 - 1}}.$$

We have the pair of results

$$\frac{d}{dx}(\cosh^{-1} x) = \frac{1}{\sqrt{x^2 - 1}}, \qquad \int \frac{1}{\sqrt{x^2 - 1}}\, dx = \cosh^{-1} x + c. \qquad (11.6)$$

Notice the difference between the integral provided by equation (11.6) and the integral that occurs in

connection with the inverse sine (equation (10.15)). The integral
$$\int \frac{1}{\sqrt{x^2-1}}\,dx$$
is used on the interval $(-\infty, -1)$ or $(1, \infty)$, while the integral
$$\int \frac{1}{\sqrt{1-x^2}}\,dx = \sin^{-1} x + c$$
is used on the interval $(-1, 1)$.

The hyperbolic sine is easier to deal with because it is 1–1, so it has an inverse, denoted $\sinh^{-1}$, with domain and range $\mathbb{R}$ (Figure 11.3).

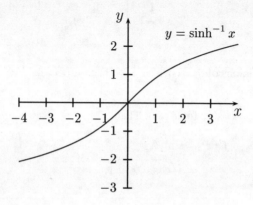

Figure 11.3: The graph of $\sinh^{-1} x$

The derivative of $\sinh^{-1} x$ is found in a similar manner to that of $\cosh x$. Let $f(x) = \sinh x$. Then
$$f'(x) = \cosh x$$
$$= \sqrt{1 + \sinh^2 x}.$$
Thus
$$\frac{d}{dx}(\sinh^{-1} x) = (f^{-1})'(x)$$
$$= \frac{1}{f'(f^{-1}x)}$$
$$= \frac{1}{\cosh(f^{-1}x)}$$
$$= \frac{1}{\sqrt{1 + \sinh^2(\sinh^{-1} x)}}$$
$$= \frac{1}{\sqrt{1 + x^2}}.$$

Thus we have the pair of results

$$\frac{d}{dx}(\sinh^{-1} x) = \frac{1}{\sqrt{1+x^2}}, \qquad \int \frac{1}{\sqrt{1+x^2}}\, dx = \sinh^{-1} x + c.$$

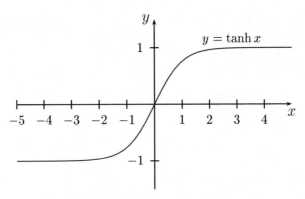

Figure 11.4: The graph of $\tanh x$

The hyperbolic tangent is also 1–1. It has domain $\mathbb{R}$ and range $(-1, 1)$ (Figure 11.4). It has an inverse, denoted $\tanh^{-1}$, with domain $(-1, 1)$ and range $\mathbb{R}$ (Figure 11.5).

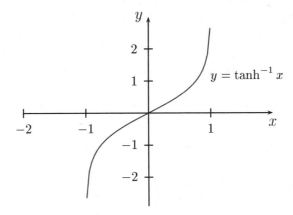

Figure 11.5: The graph of $\tanh^{-1} x$

The derivative is found in an analogous way to those of the hyperbolic sine and cosine (Exercises 11.3). We have the pair of results

$$\frac{d}{dx}(\tanh^{-1} x) = \frac{1}{1-x^2}, \qquad \int \frac{1}{1-x^2}\, dx = \tanh^{-1} x + c.$$

The inverse trigonometric functions and inverse hyperbolic functions provide a complete set of integrals for functions which are reciprocals of quadratics or reciprocals of square roots of quadratics.

234 HYPERBOLIC FUNCTIONS

This is one reason for the importance of these functions. The following list presents the results we have derived in slightly generalised form useful in applications.

$$\int \frac{1}{a^2 + x^2}\, dx = \frac{1}{a} \tan^{-1} \frac{x}{a} + c \qquad \int \frac{1}{a^2 - x^2}\, dx = \frac{1}{a} \tanh^{-1} \frac{x}{a} + c$$

$$\int \frac{1}{\sqrt{a^2 - x^2}}\, dx = \sin^{-1} \frac{x}{a} + c \qquad \int \frac{1}{\sqrt{x^2 - a^2}}\, dx = \cosh^{-1} \frac{x}{a} + c$$

$$\int \frac{1}{\sqrt{a^2 + x^2}}\, dx = \sinh^{-1} \frac{x}{a} + c$$

Table 11.1: Indefinite integrals

EXERCISES 11.3

Find the derivatives of the functions in Exercises 1–10.

1. $f(x) = \cosh^{-1} 2x$
2. $f(x) = \sinh^{-1}(x^2 + 1)$
3. $f(x) = \dfrac{1}{\tanh^{-1} x}$
4. $f(x) = \cosh^{-1} x + \sinh^{-1} x$
5. $f(x) = \sinh^{-1}(a \cosh x)$
6. $f(x) = (\sinh^{-1} x)^3$
7. $f(x) = \ln(\tanh^{-1} x)$
8. $f(x) = \tan^{-1}(\sinh x)$
9. $f(x) = \sinh^{-1} \sqrt{x^2 - 1}$
10. $f(x) = \cosh^{-1} \sqrt{x^2 + 1}$

11. Prove the following two results stated in the text:

$$\frac{d}{dx}(\tanh^{-1} x) = \frac{1}{1 - x^2}, \qquad \int \frac{1}{1 - x^2}\, dx = \tanh^{-1} x + c.$$

12. Since the hyperbolic functions can be expressed as exponentials, it is not surprising that the inverse hyperbolic functions can be expressed as logarithms. For the hyperbolic sine we have

$$\sinh^{-1} x = \ln\left(x + \sqrt{x^2 + 1}\right).$$

   (a) Prove this result by showing that the derivatives of the two sides of the equation are equal, so that the two sides differ by a constant. Show that the constant is zero.

   (b) Prove this result by letting $y = \sinh^{-1} x$ and then solving the equation $x = \sinh y$ for y in terms of x.

13. Prove the following two results

$$\cosh^{-1} x = \ln\left(x + \sqrt{x^2 - 1}\right), \qquad \tanh^{-1} x = \tfrac{1}{2} \ln\left(\frac{1 + x}{1 - x}\right).$$

CHAPTER 12
METHODS OF INTEGRATION

12.1 INTRODUCTION

In this chapter, we will look at methods of evaluating definite integrals by finding a primitive function of the integrand. This is in contrast to our discussion in Chapter 9, where the main emphasis was on using numerical methods such as Simpson's rule to evaluate definite integrals. One matter that needs to be emphasized is that a given function f, say, need not have a primitive function in terms of a given class of functions. Recall that the *elementary* functions are those functions which include algebraic formulas, roots, exponential and logarithmic functions, trigonometric functions, inverse trigonometric functions, hyperbolic functions and inverse hyperbolic functions. If f is an elementary function, we might expect to find another elementary function F such that $F' = f$ but this cannot always be done. It is known, for example, that there is no elementary function F such that $F'(x) = e^{-x^2}$. In such a case, one option is to define a new function in terms of an integral,[1] as was done in Chapter 9. For example, if we let

$$F(x) = \int_0^x e^{-t^2}\, dt,$$

then

$$F'(x) = e^{-x^2}.$$

There is no generally accepted name for the function $F(x)$ defined above. However, there is a name provided for a closely related function. We define the *generalised error function* by

$$\operatorname{Erf}(x) = \frac{2}{\sqrt{\pi}} \int_0^x e^{-t^2}\, dt.$$

We can now write

$$\int_0^x e^{-t^2}\, dt = \frac{\sqrt{\pi}}{2} \operatorname{Erf}(x),$$

so that

$$F(x) = \frac{\sqrt{\pi}}{2} \operatorname{Erf}(x).$$

The function $\operatorname{Erf}(x)$ is one whose values can be calculated by numerical integration. It is also built into software packages such as *Mathematica*. Try the *Mathematica* instruction

---
[1] Another option is to define the function by a suitable power series.

```
Plot[Erf[x],{x,-5,5}]
```
to plot its graph. Although not classed as an elementary function, it is conceptually no different to functions such as $\log x$, $\sin x$ and so on, in that there is a rule of calculation which produces an output number from a given input number. The worst that can be said about this function is that it is unfamiliar.

In principle, all definite integrals and functions defined by definite integrals can all be calculated numerically and, if need be, their graphs plotted without making use of the concept of an indefinite integral. However, this is not always necessary or efficient. A simple integral such as $\int_0^a t^2 \, dt$ can be far more efficiently evaluated by making use of primitive functions. It is also helpful to know whether the functions defined by an integral are new or whether they are known functions in another form. For example, the function

$$G(x) = \int_0^x \frac{1}{1+t^2} dt$$

is just another way of writing $\tan^{-1} x$ so that G is not a new function. For these reasons, we need methods for finding elementary functions for integrals when this is possible. There are also practical problems such as the length of a curve, areas, volumes and so on, which can be expressed as definite integrals. We will not consider such applications in detail in this book, but we will briefly summarise them below.

Areas

The area under a curve has already been dealt with in Chapter 9. If f is a given positive function, the area bounded by the graph, the x axis and the vertical lines $x = a$, $x = b$ for a and b in the domain of f is

$$A = \int_a^b f(x) \, dx.$$

Volumes of revolution

Let f be a function such that $f(x) > 0$ for $a \leqslant x \leqslant b$. If the graph of f is rotated about the x axis, then a solid is generated. The volume of this solid is

$$V = \pi \int_a^b [f(x)]^2 \, dx.$$

The length of a curve

The length of the curve $y = f(x)$ from the point $(a, f(a))$ to the point $(b, f(b))$ is

$$\ell = \int_a^b \sqrt{1 + [f'(x)]^2} \, dx.$$

Work

The work W done by a constant force on a body moving in a straight line is the product of F, the component of the force in the direction of movement, and d, the total distance moved. Thus

$$W = Fd.$$

The work done represents the increase of kinetic energy of the body. For example, for a falling body of mass m, the force is a constant mg and in falling a distance h the work done is mgh. This represents the increase in kinetic energy, which is $\frac{1}{2}mv^2$, so we have

$$mgh = \tfrac{1}{2}mv^2,$$

and hence $v = \sqrt{2gh}$. If we take the example of the *Tower of Terror*, then $h = 114\,\text{m}$, and hence $v = 44\,\text{m/sec}$.

If the force F varies, such as in a spring where $F(x) = kx$, it can be shown that

$$W = \int_a^b F(x)\,dx,$$

where $x = a$ and $x = b$ are the end points of the motion.

12.2 CALCULATION OF DEFINITE INTEGRALS

There are two methods available to compute a definite integral. We can do it numerically and (usually) approximately using a method such as Simpson's rule. However, if we can find a primitive function for the integrand then we can express the definite integral in an exact form in terms of the primitive function. We have already considered numerical evaluations in Chapter 9 and in this chapter we want to concentrate on methods for finding primitives.

There is a very useful result, related to the Fundamental Theorem of the Calculus, which we now state.

THEOREM 12.1 Evaluating definite integrals
Let f be a continuous function on the interval $[a,b]$ and let F be a primitive function for f. Then

$$\int_a^b f(x)\,dx = F(b) - F(a).$$

To prove this result, let F be any primitive of f. Then $F' = f$. Now let

$$G(x) = \int_a^x f(t)\,dt,$$

so that $G' = f$. Then F and G are both primitives of f and must differ by a constant. Thus $G(x) = F(x) + c$ for all $x \in [a,b]$. Put $x = a$. Then $0 = F(a) + c$, so that $c = -F(a)$. Hence

$$G(x) = F(x) - F(a),$$

so that

$$G(b) = F(b) - F(a),$$

which is to say

$$\int_a^b f(x)\,dx = F(b) - F(a).$$

As a matter of notation, we write $\bigl[F(x)\bigr]_a^b$ for $F(b) - F(a)$.

EXAMPLE 12.1
Since
$$\int \sin x \, dx = -\cos x + c,$$
we have
$$\int_0^\pi \sin x \, dx = \left[-\cos x \right]_0^\pi \qquad (12.1)$$
$$= -(\cos \pi - \cos 0)$$
$$= 2.$$

Notice that it is unnecessary to put the constant c in equation (12.1). If you don't understand why, try putting in the c and see what happens. □

The calculation of a definite integral by this method requires us to be able to find an indefinite integral among the known functions. There are two ways to do this:

- To make use of a table of integrals. These can range from a small table, such as Table 12.1, to large books of tables.

- If a particular indefinite integral is not on our list, we may be able to rewrite it in terms of integrals that do appear in the table. This chapter is largely about such methods.

$$\int x^n \, dx = \frac{x^{n+1}}{n+1} + c, \ n \neq -1 \qquad \int \frac{1}{a^2 + x^2} \, dx = \frac{1}{a} \tan^{-1} \frac{x}{a} + c$$

$$\int \frac{1}{x} \, dx = \log |x| + c, \ x \neq 0 \qquad \int \frac{1}{\sqrt{a^2 - x^2}} \, dx = \sin^{-1} \frac{x}{a} + c$$

$$\int e^x \, dx = e^x + c \qquad \int \frac{1}{\sqrt{a^2 + x^2}} \, dx = \sinh^{-1} \frac{x}{a} + c$$

$$\int \cos x \, dx = \sin x + c \qquad \int \frac{1}{\sqrt{x^2 - a^2}} \, dx = \cosh^{-1} \frac{x}{a} + c$$

$$\int \sin x \, dx = -\cos x + c$$

$$\int \sec^2 x \, dx = \tan x + c$$

Table 12.1: Indefinite integrals

There are two main methods of transforming integrals into a standard form. Firstly, there are methods that make use of differentiation formulas and, secondly, we can rewrite the integral by using algebraic manipulation or trigonometric identities.

EXERCISES 12.2

Evaluate the following integrals:

1. $\int_0^{\pi/2} \sin x \, dx$

2. $\int_0^{\pi/4} (2\cos x - 1) \, dx$

3. $\int_1^4 (3x^2 + 4x) \, dx$

4. $\int_0^1 \frac{1}{1+x^2} \, dx$

5. $\int_0^1 \frac{1}{\sqrt{1-x^2}} \, dx$

6. $\int_0^4 \cosh x \, dx$

7. $\int_0^\pi 2\sin x \cos x \, dx$

8. $\int_{\pi/4}^{\pi/3} (1 + \tan^2 x) \, dx$

9. $\int_1^4 \frac{1}{x^4} \, dx$

10. $\int_e^{e^3} \frac{dx}{x}$

11. $\int_0^{\log 3} e^{2x} \, dx$

12. $\int_{-1}^4 H_2(x) \, dx$,

where $H_2(x)$ is the Heaviside function defined in Exercises 2.6.

12.3 INTEGRATION BY SUBSTITUTION

This method is derived from the chain rule. Suppose F is a primitive for f. Then for any differentiable function g, we have

$$\frac{d}{dx} F(g(x)) = F'(g(x))g'(x)$$
$$= f(g(x))g'(x).$$

Thus

$$\int f(g(x))g'(x) \, dx = F(g(x)) + c. \tag{12.2}$$

This is the rule for *integration by substitution*. It is possible to use it as it stands, as is done in the following example.

EXAMPLE 12.2
Consider

$$\int 2x \sin x^2 \, dx.$$

If $g(x) = x^2$, then $g'(x) = 2x$ and the integral has the form

$$\int \sin g(x) \, g'(x) \, dx.$$

240 METHODS OF INTEGRATION

In this example, the function corresponding to f in equation (12.2) is the sine function. Thus $f(x) = \sin x$ and $F(x) = -\cos x$. So putting all of this into equation (12.2) gives

$$\int \sin g(x)\, g'(x)\, dx = -\cos g(x) + c$$
$$= -\cos x^2 + c.$$

□

You may find the above example confusing. In fact, there is no need to explicitly identify the functions F, f and g in order to use integration by substitution. It is perhaps worthwhile to use the formula once or twice in this way to convince yourself that you understand it and why integration by substitution works, but then you should approach such problems as in the next example.

EXAMPLE 12.3
In the integral

$$\int 2x \sin x^2\, dx$$

put $u = x^2$, so that

$$\frac{du}{dx} = 2x.$$

Then symbolically write $du = 2x\, dx$. In the integral we now replace x^2 by u and $2x\, dx$ by du to get

$$\int 2x \sin x^2\, dx = \int \sin u\, du$$
$$= -\cos u + c$$
$$= -\cos x^2 + c.$$

Much easier! Note that after we have made the substitution $u = x^2$, the new integral must not contain the original variable x. □

EXAMPLE 12.4
- In the integral

$$\int x\sqrt{1+x^2}\, dx,$$

put $u = x^2 + 1$ ($u = x^2$ also works), so that $du = 2x\, dx$ and hence $x\, dx = \tfrac{1}{2} du$. Then the integral becomes

$$\tfrac{1}{2}\int \sqrt{u}\, du = \tfrac{1}{2}\left(\tfrac{2}{3} u^{3/2}\right) + c$$
$$= \tfrac{1}{3}(1+x^2)^{3/2} + c.$$

- In the integral

$$\int \tan^2 x \sec^2 x\, dx,$$

put $u = \tan x$, so $du = \sec^2 x \, dx$ and the integral becomes

$$\int \tan^2 x \sec^2 x \, dx = \int u^2 \, du$$
$$= \tfrac{1}{3} u^3 + c$$
$$= \tfrac{1}{3} \tan^3 x + c.$$

□

In these examples the substitution has been given. You will improve your ability to find the correct substitution by doing the exercises—the more the better.

EXERCISES 12.3

Evaluate the following integrals:

1. $\int 2x(x^2 + 1) \, dx$

2. $\int \sin x \cos^3 x \, dx$

3. $\int 2x e^{x^2} \, dx$

4. $\int \dfrac{x}{\sqrt{x^2 + 1}} \, dx$

5. $\int \sqrt{x + 1} \, dx$

6. $\int x^2 \sqrt{x^3 + 4} \, dx$

7. $\int x^2 \cos x^3 \, dx$

8. $\int \dfrac{\sin \sqrt{x}}{\sqrt{x}} \, dx$

9. $\int \dfrac{t^{n-1}}{\sqrt{a + bt^n}} \, dt$

10. $\int (as + b)\sqrt{as^2 + 2bs + c} \, ds$

11. $\int \dfrac{\log x}{x} \, dx$

12. $\int (\sin \theta) e^{\cos \theta} \, d\theta$

13. $\int \dfrac{\cosh 2x}{1 + \sinh 2x} \, dx$

14. $\int \coth u \, du$

15. $\int \dfrac{u^2}{\sqrt{1 - u^2}} \, du$

16. $\int \dfrac{\log t^3}{t} \, dt$

17. $\int \dfrac{1}{x(\log x^2)} \, dx$

18. $\int \dfrac{\tan^{-1} x}{1 + x^2} \, dx$

19. $\int \dfrac{1}{\sqrt{1 - e^{2p}}} \, dp$

20. $\int \dfrac{\cos x}{\sqrt{1 - \sin^2 x}} \, dx$

12.4 INTEGRATION BY PARTS

Integration by parts is derived from the product rule for differentiation. If f and g are two differentiable functions, then

$$\frac{d}{dx}(f(x)g(x)) = f'(x)g(x) + f(x)g'(x).$$

If we integrate this equation we get

$$\int [f'(x)g(x) + f(x)g'(x)]\, dx = f(x)g(x),$$

which may be rearranged to give

$$\int f'(x)g(x)\, dx = f(x)g(x) - \int f(x)g'(x)\, dx. \tag{12.3}$$

There are two integrals here. If we want to evaluate the one on the left and cannot do it, then our hope is that we can do the one on the right. If neither integral can be found, then we have to try another method. The formula for integration by parts is easier to apply if we write it in the form

$$\int u\, dv = uv - \int v\, du, \tag{12.4}$$

where $u = g(x)$, $du = g'(x)\, dx$, $dv = f'(x)\, dx$ and $v = f(x)$.

EXAMPLE 12.5
To find

$$\int x \log x\, dx$$

we put $u = \log x$ and $dv = x\, dx$. We have to be guided in this choice by what we have to do with u and v. Since we need to integrate dv, it is no good taking $dv = \log x\, dx$ because we have no indefinite integral for $\log x$. So we are forced to take $dv = x\, dx$. With this choice of u and dv, we have $du = (1/x)dx$ and $v = \frac{1}{2}x^2$. Application of the integration by parts formula gives

$$\int x \log x\, dx = \frac{1}{2}x^2 \log x - \frac{1}{2}\int \frac{1}{x}x^2\, dx$$
$$= \frac{1}{2}x^2 \log x - \frac{1}{4}x^2 + c.$$

□

It is sometimes necessary to integrate by parts twice. This is particularly common in integrals involving trigonometric functions.

EXAMPLE 12.6

$$\int e^{ax} \cos bx\, dx = \frac{1}{b}e^{ax} \sin bx - \frac{a}{b}\int e^{ax} \sin bx\, dx$$
$$= \frac{1}{b}e^{ax} \sin bx - \frac{a}{b}\left(-\frac{1}{b}e^{ax} \cos bx + \frac{a}{b}\int e^{ax} \cos bx\, dx\right)$$
$$= \frac{1}{b}e^{ax} \sin bx + \frac{a}{b^2}e^{ax} \cos bx - \frac{a^2}{b^2}\int e^{ax} \cos bx\, dx.$$

The same integral appears on both sides of the equation. We combine the two integrals to get

$$\left(1 + \frac{a^2}{b^2}\right)\int e^{ax} \cos bx\, dx = \frac{1}{b}e^{ax} \sin bx + \frac{a}{b^2}e^{ax} \cos bx,$$

so that

$$\int e^{ax} \cos bx \, dx = \frac{b}{a^2+b^2} e^{ax} \sin bx + \frac{a}{a^2+b^2} e^{ax} \cos bx.$$

□

The special case $f(x) = x$ in equation (12.3) is often useful. In this case, equation (12.3) becomes

$$\int g(x) \, dx = xg(x) - \int xg'(x) \, dx. \qquad (12.5)$$

EXAMPLE 12.7

$$\int \log x \, dx = x \log x - \int x \frac{1}{x} \, dx$$
$$= x \log x - x + c.$$

□

EXERCISES 12.4

Evaluate the following integrals:

1. $\int xe^x \, dx$
2. $\int x \sin x \, dx$
3. $\int x^2 \sin x \, dx$
4. $\int \sin^{-1} x \, dx$
5. $\int \tan^{-1} x \, dx$
6. $\int x^2 \sin^{-1} x \, dx$
7. $\int e^x \sin 2x \, dx$
8. $\int x^2 \log x \, dx$
9. $\int \frac{\sqrt{x+4}}{x^2} \, dx$
10. $\int x \sec^2 x \, dx$
11. $\int (\sin^{-1} x)^2 \, dx$
12. $\int \frac{\log x}{x^2} \, dx$
13. $\int \sin x \log(\sin x) \, dx$
14. $\int x^2 (\log x)^2 \, dx$

12.5 THE METHOD OF PARTIAL FRACTIONS

The previous two methods of integration have been based on differentiation rules. The method of *partial fractions* relies on using algebraic methods to rewrite the integrand in a form which is easier to integrate.

In order to see where we are going, let's first consider an example.

EXAMPLE 12.8

It is easily checked that

$$\frac{2}{x^2 + 4x + 3} = \frac{1}{x+1} - \frac{1}{x+3}.$$

We will have to consider methods of deriving the right-hand side from the left-hand side, but we won't worry about it just yet. If we want to find the integral

$$\int \frac{2}{x^2 + 4x + 3} \, dx,$$

we can do it easily by writing

$$\int \frac{2}{x^2 + 4x + 3} \, dx = \int \left(\frac{1}{x+1} - \frac{1}{x+3} \right) dx$$
$$= \log(x+1) - \log(x+3) + c.$$

□

Partial fractions are used for functions which are quotients of two polynomials, such as

$$\frac{x+1}{x^2+4x+3}, \quad \frac{x^3+1}{x^2+4x+3}, \quad \frac{x+1}{(x+1)(x-2)(x+2)} \quad \text{or} \quad \frac{1}{(x+1)(x^2+x+1)}.$$

Functions of this type are called *rational functions*.

Let us first consider rational functions where the degree of the numerator is less than the degree of the denominator. There are two facts about polynomials which we will make use of:

- Any polynomial can be factorised into a product of linear and quadratic factors. In the general case this may be difficult and we will only deal with cases where the factorisation has already been done or is easy.

- The second fact is that once we have factorised the denominator of a rational function, then the function can be split into separate fractions, one for each factor. This is called the *partial fraction decomposition* of the function. In each fraction the degree of the numerator will be less than the degree of the denominator. An example of such a decomposition appears in Example 12.8 above.

We will only consider cases where the denominator of a rational function has factors which are all different. In this case, each linear factor $ax + b$ in the denominator will give a fraction

$$\frac{A}{ax+b}$$

in the partial fraction decomposition, where A is a constant. Each quadratic factor $ax^2 + bx + c$ gives a fraction

$$\frac{Ax+B}{ax^2+bx+c}$$

in the partial fraction decomposition, where A and B are constants.

EXAMPLE 12.9
Find the partial fraction decomposition of
$$\frac{x}{(x+1)(x+2)}.$$

Solution. The form of the partial fraction decomposition is
$$\frac{x}{(x+1)(x+2)} = \frac{A}{x+1} + \frac{B}{x+2},$$
so all we have to do is find A and B. One way to do this is to multiply through by the denominator of the left hand side. This gives
$$x = A(x+2) + B(x+1).$$
This has to hold for all values of x, so we can substitute any two values of x to get two equations in A and B. Some choices of x involve less work than others. In this case put $x = -1$ to get $-1 = A$ and then put $x = -2$ to get $-2 = -B$. Hence $A = -1$ and $B = 2$, so
$$\frac{x}{(x+1)(x+2)} = -\frac{1}{x+1} + \frac{2}{x+2}.$$
□

EXAMPLE 12.10
Find
$$\int \frac{4}{(x-1)(x^2+1)}\,dx.$$

Solution. We have
$$\frac{4}{(x-1)(x^2+1)} = \frac{A}{x-1} + \frac{Bx+C}{x^2+1}.$$
We multiply by the denominator of the left hand side to get
$$4 = A(x^2+1) + (Bx+C)(x-1).$$
Putting $x = 1$ gives $A = 2$. To find B and C we equate coefficients of like powers of x on both sides of the equation (Theorem 5.1 of Chapter 5). With practice, you should be able to do this without expanding the right hand side completely, but in this example we will give more detail. The above equation can be written
$$4 = (A+B)x^2 + (B+C)x + A - C,$$
so we have
$$A + B = 0,$$
$$B + C = 0,$$
$$A - C = 4.$$
We could solve these equations simultaneously for A, B and C, but we already know $A = 2$, so we quickly find $B = -2$ and $C = -2$.

Now we can return to the integral:

$$\int \frac{4}{(x-1)(x^2+1)} \, dx = \int \left(\frac{2}{x-1} - \frac{2x+2}{x^2+1} \right) dx$$

$$= \int \left(\frac{2}{x-1} - \frac{2x}{x^2+1} - \frac{2}{x^2+1} \right) dx$$

$$= 2\log(x-1) - \log(x^2+1) - 2\tan^{-1} x + c.$$

$\square$

We have not yet considered the case where the numerator of a rational function has degree greater than or equal to the degree of the denominator. The method here is to do a long division and then find the partial fraction decomposition of the remainder. We illustrate this in the next example.

EXAMPLE 12.11
Expand

$$\frac{2x^3 + 7x^3 + 8x + 2}{x^2 + 3x + 2}$$

in partial fractions.
Solution. On dividing the numerator by the denominator, we find

$$\frac{2x^3 + 7x^3 + 8x + 2}{x^2 + 3x + 2} = 2x + 1 + \frac{x}{x^2 + 3x + 2}$$

$$= 2x + 1 + \frac{x}{(x+1)(x+2)}$$

$$= 2x + 1 - \frac{1}{x+1} + \frac{2}{x+2},$$

where, in the last step, we have used the result in example 12.9. $\square$

The method of partial fractions can be further developed to deal with cases where higher powers of the linear or quadratic factors appear in the denominator, for example, integrals of the type

$$\int \frac{x}{(x+1)(2x+1)^3} \, dx$$

or

$$\int \frac{2x+1}{(x+1)^3(x^2+x+1)^2} \, dx.$$

We won't go into the details of how to attack these cases by hand, but we will show you how *Mathematica* can be used to find the partial fraction decompositions.

The relevant *Mathematica* instruction is Apart, so to find the partial fraction decomposition of

$$\frac{x}{(x+1)(2x+1)^3}$$

we use

```
Apart[x/((x+1)(2x+1)^3)].
```

Mathematica returns the answer

$$\frac{x}{(x+1)(2x+1)^3} = \frac{1}{1+x} - \frac{1}{(1+2x)^3} + \frac{2}{(1+2x)^2} - \frac{2}{1+2x}.$$

EXERCISES 12.5

Evaluate the following integrals:

1. $\displaystyle\int \frac{1}{(1+x)(2+x)}\,dx$

2. $\displaystyle\int \frac{1}{1-x^2}\,dx$

3. $\displaystyle\int \frac{5}{(2x+1)(x-2)}\,dx$

4. $\displaystyle\int \frac{1}{x^2-x-6}\,dx$

5. $\displaystyle\int \frac{x^2+1}{x^3+3x^2+2x}\,dx$

6. $\displaystyle\int \frac{x^2}{x^4-1}\,dx$

7. $\displaystyle\int \frac{1}{x^3+x}\,dx$

8. $\displaystyle\int \frac{2}{(x+1)(x^2+1)}\,dx$

9. $\displaystyle\int \frac{x^2}{x^4+x^2-2}\,dx$

10. $\displaystyle\int \frac{7}{(x-2)(x-5)}\,dx$

11. $\displaystyle\int \frac{x}{(x+1)(x+2)(x+3)}\,dx$

12. $\displaystyle\int \frac{1}{x(x^2+x+1)}\,dx$

Evaluate the following integrals by first making a substitution and then using partial fractions.

13. $\displaystyle\int_0^{\pi/2} \frac{\cos\theta}{\sin^2\theta + 2\sin\theta + 2}\,d\theta$

14. $\displaystyle\int \frac{\sin x}{\cos^2 x + 3\cos x + 2}\,dx$

15. $\displaystyle\int \frac{t^{2/3}+1}{t^{2/3}-1}\,dt$

16. $\displaystyle\int \frac{1}{u(1+u^{1/5})}\,du$

12.6 INTEGRALS WITH A QUADRATIC DENOMINATOR

When applying the method of partial fractions, we often end up with integrals of the form

$$\int \frac{Ax+B}{ax^2+bx+c}\,dx.$$

These can be evaluated by first writing the numerator in the form

$$\text{numerator} = k \times \text{derivative of denominator} + \text{other terms}$$

(where k is a constant) and then writing

$$\frac{\text{numerator}}{\text{denominator}} = k\frac{\text{derivative of denominator}}{\text{denominator}} + \frac{\text{other terms}}{\text{denominator}}.$$

It should now be possible to perform the integration. The procedure is best illustrated by an example.

EXAMPLE 12.12

$$\int \frac{3x+1}{x^2+2x+5}\,dx = \int \frac{\frac{3}{2}(2x+2)-2}{x^2+2x+5}\,dx$$
$$= \frac{3}{2}\int \frac{2x+2}{x^2+2x+5}\,dx - 2\int \frac{1}{x^2+2x+5}\,dx$$
$$= \frac{3}{2}\int \frac{2x+2}{x^2+2x+5}\,dx - 2\int \frac{1}{(x+1)^2+4}\,dx$$

The first integral is a logarithm, while the second may be done by letting $u = x+1$. Then we have

$$\int \frac{3x+1}{x^2+2x+5}\,dx = \frac{3}{2}\int \frac{2x+2}{x^2+2x+5}\,dx - 2\int \frac{1}{u^2+4}\,du$$
$$= \frac{3}{2}\log(x^2+2x+5) - 2\left(\frac{1}{2}\tan^{-1}\frac{u}{2}\right) + c$$
$$= \frac{3}{2}\log(x^2+2x+5) - \tan^{-1}\left(\frac{x+1}{2}\right) + c.$$

□

As well as the integrals we have dealt with here, there are many other types of integral for which a primitive function can be found. Important classes are integrals involving square roots of quadratics, such as

$$\int \sqrt{x^2+1}\,dx,$$

or those involving powers of trigonometric functions such as

$$\int \sec^3 x\,dx.$$

Usually *Mathematica* can handle such integrals without any problems. There are two *Mathematica* instructions for integration, namely Integrate and NIntegrate. The first instruction expects *Mathematica* to produce an exact result, usually using an indefinite integral. If it cannot do this then the second instruction will evaluate a definite integral numerically. We use Integrate as follows. For indefinite integrals we state the function to be integrated and the variable of integration:

Integrate[f[x],x]

For definite integrals, we also have to state the range of integration:

Integrate[f[x],{x,a,b}]

The instruction NIntegrate is used for definite integrals only, so we use the form

NIntegrate[f[x],{x,a,b}]

EXAMPLE 12.13
Evaluate the integrals $\int \sec^3 x\,dx$ and $\int_0^5 e^{-t^2}\,dt$.

- To evaluate $\int \sec^3 x \, dx$ use

$$\texttt{Integrate[(Sec[x])\^{}3,x]}.$$

The result given by *Mathematica* is, after some rearragement of the terms,

$$\int \sec^3 x \, dx$$
$$= \frac{1}{2}\bigg[-\log\left(\cos\left(\frac{x}{2}\right) - \sin\left(\frac{x}{2}\right)\right) + \log\left(\cos\left(\frac{x}{2}\right) + \sin\left(\frac{x}{2}\right)\right)$$
$$+ \sec(x)\,\tan(x)\bigg].$$

- To evaluate

$$\int_0^5 e^{-t^2} \, dt,$$

use

$$\texttt{NIntegrate[E\^{}(-t\^{}2),\{t,0,5\}]}.$$

The result is

$$\int_0^5 e^{-t^2} \, dt = 0.886227.$$

□

12.7 CONCLUDING REMARKS

There are two main reasons for calculating definite integrals in terms of indefinite integrals.

- Generally speaking, it is easier to evaluate definite integrals exactly if we can find a primitive of the integrand. Many quantities (area, volume, work) can be expressed as definite integrals. If we can compute them exactly by using indefinite integrals, then we do so. Otherwise we have to use numerical methods to calculate them approximately.

- Many problems require us to define functions in terms of definite integrals and it is often useful to know whether this defines a new function or is just another way of expressing a known function. For example, the function

$$F(x) = \int_0^x e^{-t^2} \, dt$$

cannot be calculated in terms of the functions we have met earlier in this book. This is a fact which can be proved and it is not just that we are not clever enough to think of a way to do it. Consequently, F is a new function. On the other hand, the function

$$G(x) = \int_0^x \frac{1}{1+t^2} \, dt$$

is not a new function. We know that $G(x) = \tan^{-1} x$.

EXERCISES 12.7

Evaluate the following integrals:

1. $\int \dfrac{1}{x^2 + 8x + 20}\,dx$

2. $\int \dfrac{x}{x^2 + x + 1}\,dx$

3. $\int \dfrac{2x + 1}{4x^2 + 12x - 7}\,dx$

4. $\int \dfrac{e^x}{e^{2x} + e^x + 1}\,dx$

5. $\int \dfrac{\cos x}{\sin^2 x - 2\sin x + 5}\,dx$

6. $\int \dfrac{2x + 3}{9x^2 + 6x + 5}\,dx$

7. $\int \dfrac{x + 3}{x^2 + 4x + 5}\,dx$

8. $\int \dfrac{3x + 1}{x^2 + x + 1}\,dx$

9. $\int \dfrac{x^2 + 1}{x^3 + x^2 + x}\,dx$

10. $\int \dfrac{\sec^2 x}{\sec^2 x + 4\tan x + 4}\,dx$

CHAPTER 13
A NONLINEAR DIFFERENTIAL EQUATION

In this chapter, we will discuss some advanced topics which arise by considering the motion around the curved part of the track in the *Tower of Terror*. We will find that the differential equation which governs this motion is also applicable to the case of the simple pendulum, which we will also take as an example in our discussion.

In Chapter 9 we derived the differential equation

$$mr\theta'' + mg \sin \theta = 0$$

to describe the motion around the curved part of the track in the *Tower of Terror*, where r is the radius of the circular part of the track. (We can remove the mass m from the equation as we did in Chapter 9, but the solution will have more physical meaning if we leave it in place.) A simple pendulum consists of a mass suspended from a fixed point by a light inelastic string (Figure 13.1). Since it consists of a mass constrained to move on a circular path influenced only by gravity, the equation for the *Tower of Terror* will apply equally to the simple pendulum.

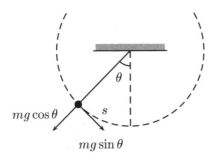

Figure 13.1: The simple pendulum

We have seen two methods for constructing solutions of differential equations, namely power series and integrals. We used these methods to define logarithmic, exponential, hyperbolic and trigonometric functions. These functions, together with their inverses and all possible compositions of them are called the *elementary functions*. With the elementary functions at our disposal, we can solve a large number of differential equations, as we did, for example, in the case of forced damped oscillations. There are, however, differential equations which do not have elementary functions as their solutions. This is simply a fact of life which we have to accept. Such problems are inherently difficult because their solutions are functions which are more complicated than elementary functions. The differential equation we are now considering is such an example.

13.1 THE ENERGY EQUATION

We recall from Chapter 8 that a differential equation is *linear* if it has the form

$$a(t)\theta'' + b(t)\theta' + c(t)\theta = F(t).$$

In a linear differential equation, the dependent variable and its derivatives occur to the first power only. The equation

$$mr\theta'' + mg\sin\theta = 0 \qquad (13.1)$$

is not linear because of the $\sin\theta$ term. It is called a *nonlinear* differential equation and the presence of the $\sin\theta$ makes it impossible to obtain a power series solution in the way we have done for ealier equations. The equation can be solved by integration, but the solution is in terms of new non-elementary functions.

As in Section 10.3, we begin by multiplying the equation by θ' to get

$$mr\theta'\theta'' + mg\theta'\sin\theta = 0$$

and then multiply by r to get

$$mr^2\theta'\theta'' + mgr\theta'\sin\theta = 0.$$

Now we express the left hand side as a time derivative. As before,

$$mr^2\theta'\theta'' = \frac{d}{dt}\left(\frac{1}{2}mr^2\theta'^2\right),$$

and by the chain rule we have

$$mgr\theta'\sin\theta = \frac{d}{d\theta}(-mgr\cos\theta)\frac{d\theta}{dt} = \frac{d}{dt}(-mgr\cos\theta).$$

Hence the differential equation becomes

$$\frac{d}{dt}\left(\frac{1}{2}mr^2\theta'^2 - mgr\cos\theta\right) = 0,$$

and so

$$\frac{1}{2}mr^2\theta'^2 - mgr\cos\theta = c. \qquad (13.2)$$

As yet we have not imposed any initial conditions. Let's suppose the body starts at the lowest point with velocity v_0. Recall that $s = r\theta$ and so

$$v(t) = \frac{ds}{dt} = r\frac{d\theta}{dt}.$$

The initial conditions then become $\theta(0) = 0$ and $\theta'(0) = v_0/r$. Using these initial conditions in equation (13.2) gives

$$c = \frac{1}{2}mr^2\frac{v_0^2}{r^2} - mgr$$
$$= \frac{1}{2}mv_0^2 - mgr.$$

Hence

$$\frac{1}{2}mr^2\theta'^2 - mgr\cos\theta = \frac{1}{2}mv_0^2 - mgr,$$

or

$$\frac{1}{2}mr^2\theta'^2 + mgr(1 - \cos\theta) = E, \quad (13.3)$$

where

$$E = \frac{1}{2}mv_0^2.$$

This result is an equation expressing conservation of energy. The expression $\frac{1}{2}mv_0^2$ is the initial kinetic energy (KE) of the body, which occurs at the lowest point. As it travels up the arc, the body's velocity is given by $r\theta'(t)$ and so $\frac{1}{2}mr^2(\theta'(t))^2$ is its kinetic energy at time t. The second term on the left hand side is the potential energy (PE) and is the gravitational energy gained by the body rising above the initial level. It is zero at the lowest point where $\cos\theta = 1$. This energy would be released if the body fell back to the original level. Thus the above equation says that

$$(\text{KE} + \text{PE})\big|_{\text{time } t} = (\text{KE} + \text{PE})\big|_{\text{initially}}.$$

From the energy equation (13.3) we obtain

$$\theta'^2 = \frac{2E}{mr^2}\left(1 - \frac{2mgr}{E}\sin^2\left(\frac{\theta}{2}\right)\right). \quad (13.4)$$

The next step depends on the values of the constants and in particular on the relation between E and $2mgr$. The reason for this can be seen physically if we imagine the body to be going around a circular track. We assume that it is constrained to the track and cannot drop off the top portion. At the top of the circle, the height is $2r$ and the PE is $2mgr$. If $E > 2mgr$, then the body will have positive KE when it reaches the top and so it will rotate. If $E < 2mgr$, then it is impossible for the body to reach the top—it cannot have negative KE—and so it will fall backward and oscillate. If $E = 2mgr$, the body will just reach the top and then stop—a very unlikely outcome in practice. The solutions are different in each case. Mathematically this corresponds to whether the term

$$1 - \frac{2mgr}{E}\sin^2\left(\frac{\theta}{2}\right)$$

can be zero or not. If $E < 2mgr$, then it can be zero and this will allow θ' to be zero, and the times at which this occurs will be when the motion reverses in the oscillating mode.

For the *Tower of Terror* we have $v = 160$ km/hr or 44 m/s and shall assume $r = 25$ m. Then

$$\frac{2mgr}{E} = \frac{2mgr}{\frac{1}{2}mv^2} = \frac{4gr}{v^2} \doteq 0.5,$$

so that we definitely have $E > 2mgr$.

13.1.1 The case $E > 2mgr$

Let $k^2 = 2mgr/E$ and $\omega^2 = 2E/mr^2$. Then equation (13.4) can be written as

$$\theta'^2 = \omega^2 \left(1 - k^2 \sin^2(\theta/2)\right),$$

and so

$$\frac{1}{\sqrt{1 - k^2 \sin^2(\theta/2)}} \frac{d\theta}{dt} = \omega.$$

For convenience, we substitute $\alpha = \theta/2$. Then the last equation becomes

$$\frac{1}{\sqrt{1 - k^2 \sin^2 \alpha}} \frac{d\alpha}{dt} = \frac{\omega}{2}. \tag{13.5}$$

In order to integrate the left hand side, we need a function F whose derivative is given by

$$F'(x) = \frac{1}{\sqrt{1 - k^2 \sin^2 x}}.$$

There is no elementary function with this property and so we have to define the new function F by letting

$$F(x) = \int_0^x \frac{1}{\sqrt{1 - k^2 \sin^2 y}} \, dy. \tag{13.6}$$

The values of this function depend on the constant k. In fact, there is not just one function, but rather there is a different function for each value of k. If we want to emphasize the value of k, we write $F(x; k^2)$ or $F(x; \beta)$, where $\beta = k^2$. This function is called an *elliptic integral of the first kind*. We can express equation (13.5) in terms of F by writing

$$\frac{d}{dt}(F(\alpha)) = \frac{\omega}{2},$$

and so we get

$$F(\alpha) = \frac{\omega t}{2} + c.$$

The initial condition $\theta(0) = 0$ gives $\alpha(0) = 0$. Consequently $F(\alpha(0)) = 0$ and hence $c = 0$, so that

$$F(\alpha) = \frac{\omega t}{2}. \tag{13.7}$$

A symbolic manipulation package such as *Mathematica* has many functions built into it besides the elementary functions. Among these are the elliptic functions. Very few calculators have built-in elliptic functions, but they easily could if there were sufficient demand. *Mathematica* can be used to calculate the function F; the notation used is `EllipticF[x,b]`. Here is a plot of $F(x; 1)$, using the *Mathematica* instruction

```
Plot[EllipticF[x,1],{x,-1.5,1.5},PlotStyle->RGBColor[1,0,0]]
```

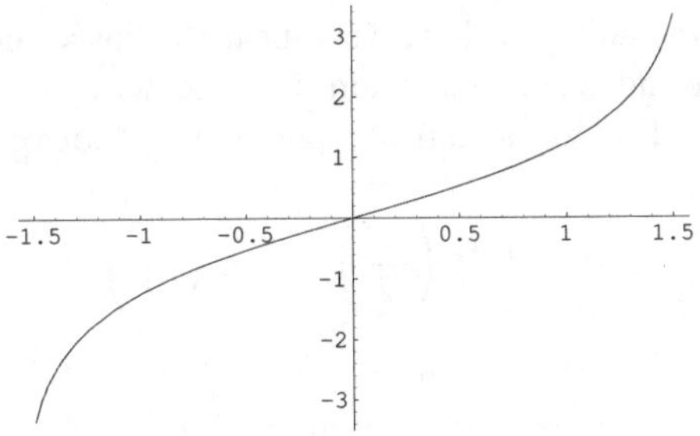

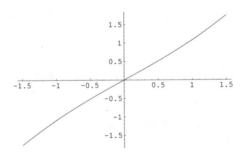

Figure 13.2: Graph of $F(x; 1)$

Note the similarity of the definition of F to that of the inverse sine function in Chapter 10 and the similarity of their graphs. If the graph of $F(y; k^2)$ is plotted for $k = \frac{1}{2}$, then the graph is almost a straight line (Figure 13.3). In fact, for small k,

$$\frac{1}{\sqrt{1-k^2\sin^2 y}} \approx 1.$$

Hence, for small k,

$$F(x; k^2) \approx \int_0^x dy = x.$$

Figure 13.3: Graph of $F(x; 0.5)$

EXAMPLE 13.1

Use *Mathematica* to evaluate $F(1; 0.5)$ in two ways.

Solution. We first use the `NIntegrate` instruction of *Mathematica* and equation (13.6) to evaluate $F(1, 0.5)$. The full *Mathematica* instruction is

`NIntegrate[1/Sqrt[1-0.5 Sin[t]^2],{t,0,1}]`

and gives the answer 1.08322. Next, we use *Mathematica*'s built-in function `EllipticF[1,0.5]` to get the same answer.

□

By the fundamental theorem of the calculus, the function F has derivative

$$F'(y) = \frac{1}{\sqrt{1-k^2\sin^2 y}} > 0,$$

so F is an increasing function and hence 1–1. This means F has an inverse. This inverse function is called the Jacobi amplitude and is denoted by am. It is another nonelementary function, not familiar from elementary calculus. The *Mathematica* expression is `JacobiAmplitude`. Using am in equation (13.7), we have

$$\alpha = F^{-1}\left(\frac{\omega t}{2}\right) = \operatorname{am}\left(\frac{\omega t}{2}\right)$$

and so

$$\theta = 2\alpha = 2\operatorname{am}\left(\frac{\omega t}{2}\right). \tag{13.8}$$

In principle, we can compute function values of am from those of F by using, for example, the Newton–Raphson process. In fact, the function am is built into *Mathematica* and we can plot its graph as in Figure 13.4, using the *Mathematica* instruction

```
Plot[JacobiAmplitude[x,1],{x,-1.5,1.5}].
```

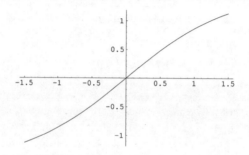

Figure 13.4: The Jacobi Amplitude function.

EXAMPLE 13.2

Find the position of the car in the *Tower of Terror* after 0.25 s. Find also the time taken to traverse the total curved portion of the path.

Solution. We have

$$\omega = \sqrt{\frac{2E}{mr^2}} = \sqrt{\frac{mv^2}{mr^2}} = \frac{v}{r} \simeq 1.67,$$

where $v \simeq 44$ m/s and $r \simeq 25$ m. We also found that $k^2 \doteq 0.5$ (page 253). Hence, using equation (13.8), we have

$$\theta(0.25) = 2\operatorname{am}\left(\frac{1.67 \times 0.25}{2}; 0.5\right) \simeq 0.42 \text{ radians} = 23 \text{ deg}.$$

To find the time taken to traverse the curved portion of the path, we need t when $\theta = \pi/2$. Then

$$\frac{\pi}{2} = 2\operatorname{am}(0.88t),$$

and so

$$t = \frac{1}{0.88}F\left(\frac{\pi}{4}\right) = 0.94 \text{ s}.$$

It is important to appreciate the similarity between the method of solution of this problem and the way we solved the oscillation problem on page 215 and the air pressure problem on page 210. There is a systematic approach here which is applicable to more problems than just these three, and it is worth devoting some effort to understanding it. In both cases the differential equation leads us to define a function with a given derivative by means of an integral. The solution is then obtained by using the inverse of this function. The difference is in the definition of the new function. In this problem it is

$$F(x) = \int_0^x \frac{dy}{\sqrt{1 - k^2 \sin^2 y}},$$

whereas in the air pressure problem it was

$$F(x) = \int_1^x \frac{1}{y} dy$$

and in the oscillation problem it was

$$F(x) = \int_0^x \frac{dy}{\sqrt{1 - y^2}}.$$

13.1.2 The case $E \leqslant 2mgr$

The cases $E < 2mgr$ and $E = 2mgr$ do not arise in the *Tower of Terror*, but they do arise in other situations—for example in the simple pendulum. There are also amusement park rides where a train oscillates or rotates on the interior of a circular track. Before continuing, let's make a few remarks about the simple pendulum. In spite of the name, you will find that a full analysis of the simple pendulum is anything but simple. In fact, the name is only there to distinguish it from the *compound pendulum* which we shall not consider. A *simple pendulum* consists of a mass of negligible dimensions suspended from a fixed point by an inelastic rigid rod of negligible mass. We require a rigid rod rather than a string, so that the pendulum is able to oscillate through any angle, or even rotate completely around its point of suspension. The motion of the pendulum depends on its velocity, and hence energy, at A and it might oscillate, rotate or in the transitional case, simply move to the top of the circle and remain there.

To solve the differential equation of motion for the case $E \leqslant 2mgr$, we use similar methods to the previous case $E > 2mgr$, but the functions which arise are different. We will simply quote the results and use *Mathematica* to plot the relevant functions.

In the case of low energy ($E < 2mgr$), the system oscillates and the solution is given by

$$\theta(t) = 2 \sin^{-1}(k \operatorname{sn} \omega t), \tag{13.9}$$

where $k^2 = E/(2mgr)$, $\omega^2 = g/r$ and sn is a function called the Jacobi elliptic sine function. Its definition is in terms of the Jacobi amplitude am and is given by

$$\operatorname{sn}(x; k^2) = \sin(\operatorname{am}(x; k^2)).$$

It is contained in *Mathematica*, where it is denoted by `JacobiSN`. Its values can be calculated and its graph plotted as for any other function.

For various values of k, we can use equation (13.9) to get graphs of the solution θ as a function of time. The exact nature of the solution will depend on the initial conditions, namely that at the lowest point, the body has velocity v. We expect that the amplitude of the oscillations will increase as v increases, and this is indeed what happens. Before plotting these graphs, it is interesting to look at a common approximation that is normally made for the simple pendulum. The equation of motion is

$$mr\theta'' + mg\sin\theta = 0,$$

which we write as

$$\theta'' + \frac{g}{r}\sin\theta = 0.$$

If θ is small, then $\sin\theta \simeq \theta$ and the equation is approximated by

$$\theta'' + \frac{g}{r}\theta = 0,$$

with initial conditions $\theta(0) = 0$ and $\theta'(0) = v/r$. This is now a linear equation which we have seen before. Its solution is

$$\theta(t) = \frac{v}{r\omega}\sin\omega t. \tag{13.10}$$

How good is the approximation (13.10)? We have drawn three graphs below for initial velocities of 8.3 m/s, 25 m/s and 30.5 m/s. In each graph, the solid line is the solution (13.9), while the dotted line is the approximate solution (13.10). The approximation is excellent for the first case, but gets worse as we increase the initial velocity.

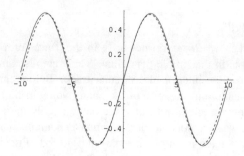

Figure 13.5: The case $v = 8.3$ m/s, $k^2 = 0.07$, $\omega = 0.63$.

Here, and in the next two graphs, we have used $r = 25$, $g = 9.81$ and

$$\omega = \sqrt{\frac{g}{r}} = 0.63, \quad k^2 = \frac{4gr}{v^2}.$$

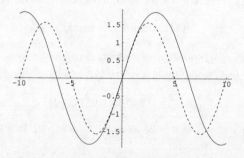

Figure 13.6: The case $v = 25$ m/s, $k^2 = 0.64$, $\omega = 0.63$.

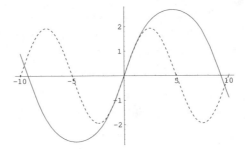

Figure 13.7: The case $v = 30.5$ m/s, $k^2 = 0.95$, $\omega = 0.63$.

For completeness, we mention the case when $E = 2mgr$. Here the body has just enough energy to reach the top of the circle, but not enough to go past it and rotate. In this case, $k = 1$ and so F is the function

$$F(x) = \int_0^x \frac{dy}{\sqrt{1 - \sin^2 y}}.$$

This function is an elementary function and can be calculated using suitable methods of integration. The elementary function solution we obtain is

$$\theta(t) = 2\sin^{-1}(\tanh \omega t),$$

where $\omega = \sqrt{g/r}$. The graph is plotted in Figure 13.8. Notice that $\theta = \pi$ and $\theta = -\pi$ are horizontal asymptotes, indicating that the body takes forever to reach the top.

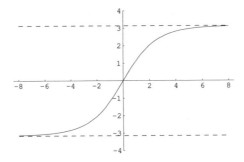

Figure 13.8: The critical case $E = 2mgr$.

13.2 CONCLUSION

The functions we have used in dealing with this problem in circular motion may be unfamiliar, but the principles used for defining them by integrals and inverse functions are exactly the same as those for trigonometric and inverse trigonometric functions that arose from the problem on the Anzac Bridge. As practical problems become more complicated, the functions which arise as solutions also become more complicated. This is simply a fact of life. Fortunately, much of the drudgery is now taken out of the calculations by tools such as *Mathematica*. Nobody has to do heroic calculations any more—computers are far quicker and more accurate—but we do need to understand the principles behind these calculations.

ANSWERS

Answers 1.3

1. Automobile braking systems
2. 19620 N/m², 20209 N/m²
3. 10200 kg/m²

Answers 1.6

1. A reasonable domain for h would be $0 \leqslant h \leqslant$ maximum sea depth. The maximum sea depth is about 12 km, but it is unlikely that any normal submarine could dive to such a depth.
2. There is only one value of h for each value of t, but (usually) two values of t for each value of h. A reasonable domain for t is all t for which $0 \leqslant t \leqslant 9$.
3. Either variable could be taken as the independent variable. A prediction for the population in 2000 is 20 million.

Answers 2.1

1. $6, 22$ 2. $\frac{1}{13}$ 3. $0, 1, 1$ 4. $1/(3-t^2)$ 5. $5/4$ 6. 3 7. 35.508267
8. $x^2 + 3, 3 + 1/x^2, x^2 + 2x + 4$ 9. $1/3$
10. $\dfrac{1-x}{x^2+1}, \dfrac{x(1+x)}{x^2+1}, \dfrac{x+2}{x^2+2x+2}$ 11. $1 - x, 1 + 1/x, x + 2$

12. $f(-x) = \begin{cases} -4x + 1, & x \geqslant -2 \\ -2x^3 - 7, & x < -2 \end{cases}$

$f(1/x) = \begin{cases} 4/x + 1, & x \geqslant 1/2 \text{ or } x < 0 \\ 2(x)^{-3} - 7, & 0 < x < 1/2 \end{cases}$

$f(x+1) = \begin{cases} 4x + 5, & x \leqslant 1 \\ 2(x+1)^3 - 7, & x > 1 \end{cases}$

13. (a) 45 m/s, (b) 88.2 m ($BD = 103.2$ m)
15. $P(107) = 24$ means that there were 24 million people in the city in 1983.
16. The values $f(x)$ and $g(x)$ are equal at all points $x \in \mathbb{R}$, so $f = g$. The points $x = \pm 1$ are not in the domain of h, but they are in the domain of f and g, so $f \neq h, g \neq h$.

17. $d = \frac{1}{3}\sqrt{10x_0^2 - 2x_0 + 1}$

18. $g(x) = \begin{cases} x - 1, & x \geq 1 \\ 1 - x, & x \leq 1 \end{cases}$

19. (a) $N = kP$, where N is the number of vehicles, P is the population and k is some constant.
    (b) $KE = kv^2$, where KE is the kinetic energy, v is the velocity and k is some constant.
    (c) $S = kr^2$, where S is the surface area, r is the radius and k is some constant.
    (d) $F = G\dfrac{Mm}{r^2}$, where F is the force and G is the universal gravitational constant.

Answers 2.2

1. $\mathbb{R}$; $[0, \infty)$
2. $\mathbb{R} \setminus \{1\}$; $\mathbb{R} \setminus \{0\}$
3. $(-\infty, \frac{1}{2}]$; $[0, \infty)$
4. $\mathbb{R} \setminus \{-1, 3\}$; $\mathbb{R}$
5. $[0, 1)$; $[0, 1)$
6. $[0, \infty)$; $[\sqrt{3}, \infty)$
7. $f(C) = C^2/(4\pi)$
8. $f(a) = a^2\sqrt{3}/4$
9. $5 < x \leq 8$
10. $1 < x < \infty$ (or $x > 1$)
11. $1 < x < 2$
12. $1 < x \leq 9$ or $x \geq 10$
13. $2 \leq x < 6$
14. $-\infty < x < \infty$ (or $x \in \mathbb{R}$)
15. $x \in [2, 3)$
16. $x \in (\pi\infty)$
17. $x \in (-\infty, \infty)$
18. $x \in [12, \infty)$
19. $x \in (-7, 1)$
20. $x \in (-\infty, -2] \cup [2, \infty)$
21. No such x exists.

Answers 2.3

2. Use the ranges $0 \leq x \leq 2, 0 \leq x \leq 10, 0 \leq x \leq 39$.
6. $x \approx 1.4\sqrt{r}$. This is easier to see if you take the scale on the x axis as multiples of $\sqrt{r}$.

Answers 2.4

1. $\mathbb{R}$; $[-3, \infty)$
2. $x \neq 2, -2$; $\mathbb{R}$
3. $t \geq 1$ and $t \leq -1$; $[0, \infty)$
4. $\mathbb{R}$; $[-3, \infty)$
5. All $x \neq 0$; All $|x| \geq 2$
6. $\mathbb{R}$; $\mathbb{R}$
7. $f(0) = 1$, $f(0.5) = 0.607$, $f(1) = 0.368$, $f(2) = 0.135$
8. As one example $f(x) = \begin{cases} -\sqrt{1-x^2}, & -1 \leq x \leq 0 \\ \sqrt{1-x^2}, & 0 < x \leq 1 \end{cases}$

Answers 3.1

1.

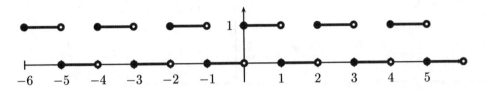

Answers 3.2

1. n^2
2. $\dfrac{1}{n}$
3. $\dfrac{1}{n(n+1)}$
4. $\dfrac{(-1)^n n}{(n+1)^2}$
5. $\dfrac{(-1)^{n+1} n}{n+1}$
6. 1
7. $\dfrac{1}{2}, \dfrac{1}{4}, \dfrac{1}{8}, \dfrac{1}{16}$
8. $\dfrac{1}{8}, \dfrac{1}{15}, \dfrac{1}{24}, \dfrac{1}{35}$
9. $-\dfrac{2}{1}, \dfrac{3}{2}, -\dfrac{4}{3}, \dfrac{5}{4}$
10. $\dfrac{2}{3}, \dfrac{8}{9}, \dfrac{26}{27}, \dfrac{80}{81}$
11. $\dfrac{2}{1}, \dfrac{1}{2}, \dfrac{4}{3}, \dfrac{3}{4}$
12. $\sqrt{2}, \sqrt{3}, \sqrt{4}, \sqrt{5}$
13. Convergent, limit 1
14. Divergent
15. Conv, 0
16. Conv, 3
17. Conv, $\sqrt{3}$
18. Conv, 2
19. Div
20. Conv, 0
21. $f(1) = 2$
22. not possible
23. $f(1) = -3$
24. $f(1) = 3$
25. $f(1) = \dfrac{1}{4}$
26. The function is discontinuous at $t = 3, 4, 5$.

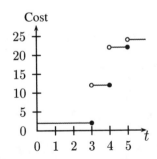

27. $f\{2 + 1/n\}$ converges to 3, $f\{2 - 1/n\}$ converges to 4.

28. The graphs of $H_1(t)$ and $P_{0.5,1}(t)$ are shown in the diagram.

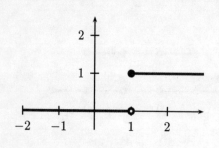
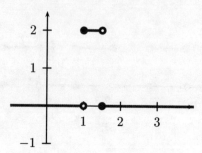

29. The diagram shows $H_{-1}(-t)$.

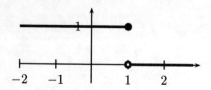

31. $28 + \dfrac{10}{n} + \dfrac{1}{n^2}$ 32. $28 - \dfrac{10}{n} + \dfrac{1}{n^2}$

33. $1 - \dfrac{3}{n} + \dfrac{3}{n^2} - \dfrac{1}{n^3}$ 34. x_n^2

Answers 4.1

1. 4   2. 28   3. $1/2\sqrt{3}$ 4. $-1/25$ 5. 36   6. $-1/9$
7. $f'(x) = 0$ 8. $f'(x) = 1$ 9. $f'(x) = 3$
10. $f'(x) = -1/(x+2)^2$ 11. $g'(x) = 4x$ 12. $h'(x) = 3x^2 + 4$

13. The function f' is differentiable at all points except $x = 0$. The graph of f' is shown in the diagram:

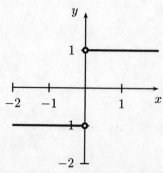

14. Both are differentiable at $x = 0$ (and at all other points).

15.

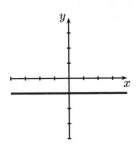

16.

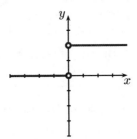

17.

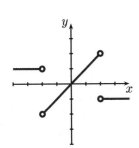

18.

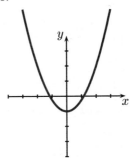

19. (a) The quotient represents the average velocity during the time interval $[x, x+h]$.

(b) The quotient represents the approximate change in the number of people per unit of time during the time interval $[x, x+h]$.

(c) If the number of units produced increases by h, then the quotient represents the approximate change in the cost per unit.

Answers 4.2

1. $f'(x) = 2x + 3$
2. $f'(x) = 6x^2 + 6x$
3. $f'(x) = -1/(1+x)^2$
4. $f'(x) = 6x^2 + 2x - 2$
5. $f'(x) = (x^2 - 2x - 2)/(x-1)^2$
6. $f'(x) = 48x^5 + 20x^4 + 4x + 1$

Answers 4.3

1. $v_{av} = \dfrac{(t_0 + h)^2 - t_0^2}{h} = 2t_0 + h$.
2. 7, 6.1, 6.01, 6.001
3. $v_{inst} = 6$
5. 39.2 m/s
6. After 1 s, 5 s; -12 m/s^2, 12 m/s^2
7. $2\pi r$, 6π
8. $A'(x) = x$, 3
9. $A'(t) = 8\pi t$, 40π m/s^2
13. 9.7 m/s, 14.4 m/s, 2.4 m/s^2
14. $0 \leqslant s \leqslant$ length of the river $\approx 60\,km$, $\quad f'(s) < 0$

15. Approximately 0.08% per hour; approximately 0.025% per hour; before 12 min and after 3.6 hours. Note that if a person was involved in a serious accident after 5 minutes and then breathalysed by the police half an hour later, that person would probably have a hard time convincing the police that he or she was under the legal limit at the time of the accident.

16. $W' = -2GMm/r^3 < 0$. The weight decreases as the ship moves away from the earth.

Answers 5.1

1. It is best to use central differences in this problem.
   (a) $\alpha'(0.1) = 0.06$, $\alpha'(0.2) = 0.12$, $\alpha'(0.3) = 0.18$, $\alpha'(0.4) = 0.24$, $\alpha'(0.5) = 0.3$
   (b) The graph is the straight line of slope 0.6 and passing through the origin.
   (c) $\alpha'(t) = 0.6t$.
   (d) $\alpha(t) = 0.3t^3 + 0.1$, $\quad \alpha(0.9) = 0.343$

2. Again, we use central differences.
   (a) $v'(3) \approx 5.5$, $v'(6) \approx 3.83$, $v'(9) \approx 3.0$
   (b) The acceleration is *decreasing*, so $v'(t)$ is negative.
   (c) 66 m/s

Answers 5.2

1. $a = 0, b = 1, c = 3, d = 0$
2. $a = b = c = 0$
3. $a = 1, b = c = 0$
4. $a = 3 = c, b = 1$
6. $y(t) = \dfrac{t^3}{3} + t + 3$
7. $h(t) = \dfrac{t^4}{4} + \dfrac{2t^3}{3} + 3t$
8. $v(t) = \dfrac{t^3}{3} + t^2 + 4t + 2$

9. The equation of motion is $s = gt^2/2$. The seventh observer has made an error. The reading should be 1.99, not 1.91.

Answers 5.3

2. The ball takes longer to fall than to rise. [Use energy considerations.]
3. 58.8 m
5. $1 : \sqrt{2}$
6. Differentiating your answer should give the differential equation you started with.

Answers 5.5

1. Increases on $(-\infty, -1)$ and $(1, \infty)$, decreases on $(-1, 1)$
2. Increases on $(-\dfrac{3}{4}, \infty)$, decreases on $(-\infty, -\dfrac{3}{4})$
3. Increases on $(-\dfrac{1}{2}, \infty)$, decreases on $(-\infty, -\dfrac{1}{2})$

4. Increases on $(-\frac{7}{2}, \infty)$, decreases on $(-\infty, -\frac{7}{2})$
5. Global minimum of -8 at $x = 0$, no local maximum
6. Local maximum of 0 at $x = 0$, local minimum of -81 at $x = \pm 3$
7. Local maximum of 16 at $x = 0$, global minima of 0 at $x = \pm 4$
8. No local extreme values, no stationary points
9. Global minimum of -3 at $x = 2$, global maximum of 1 at $x = 0$, local maximum of -2 at $x = 3$
10. The rectangle should be a square of side 6 cm.
11. $10\sqrt{6} \times 10\sqrt{3}$
12. Width of base: 11 cm, height of sides: 5.5 cm

Answers 6.1

1. $\dfrac{d\rho}{dh} = -0.13\rho$

Answers 6.2

1. 16
2. 15
3. 39
4. $\frac{2}{3}$
5. -6
6. $\frac{1}{7} + \frac{1}{9} + \frac{1}{11}$
7. $\dfrac{1}{2}$
8. 6
9. $\dfrac{4}{5}$
10. 4
11. $\dfrac{1}{192}$
12. $-\dfrac{1}{5}$
13. $\sum\limits_{k=1}^{\infty} k$
14. $\sum\limits_{k=1}^{\infty} (2k-1)$
15. $\sum\limits_{k=1}^{7} \dfrac{1}{2k+1}$
16. $\sum\limits_{k=1}^{n} 1$
17. $\sum\limits_{k=1}^{\infty} \dfrac{1}{k(k+1)}$
18. $-\sum\limits_{k=1}^{5} (k+2)(-x)^k$
19. 9
20. -1
21. -17
22. $3/5$
23. Use the fact that $1 + 2 + 3 + \cdots + n$ is an arithmetic progression.
24. $S_k = \dfrac{1}{\sqrt{2}} - \dfrac{1}{\sqrt{k+2}}$, $S_\infty = \dfrac{1}{\sqrt{2}}$
25. $S_k = \dfrac{k}{k+1}$, $S_\infty = 1$
26. $S_k = \dfrac{k}{2(k+2)}$, $S_\infty = \dfrac{1}{2}$
27. $S_k = \dfrac{6k}{1+2k}$, $S_\infty = 3$

Answers 6.3

1. $3.7407, C$
2. $1626, D$
3. $0.5917, C$
4. $7637.625, C$
5. $1.1767, D$
6. $845.537, D$
7. $0.1260, D$
8. $0.7167, C$
9. $5, D$
10. $3.2963, C$
12. 3.049. This is about 3% accuracy.
13. $\exp(0.4) \approx 1.49173$, error ≤ 0.000093

14. $\exp(0.3) \approx 1.349858$, error $\leqslant 1.3 \times 10^{-6}$

Answers 6.4

1. ∞
2. 3
3. $1/\sqrt{2}$
4. 2
5. 1
6. ∞
7. $\sqrt{3}$
8. 0
9. ∞
10. ∞
11. $0 < x < 2$
12. $-7 < x < 3$
13. $-\infty < x < \infty$
14. $-2 < x < 0$
15. $-\infty < x < \infty$
16. The terms in the sum on the right-hand side are smaller for a given value of n, so that we expect the right-hand side to converge more quickly.

Answers 6.5

1. Radius of convergence $= 1$, $f'(x) = 1 + 4x + 9x^2 + 16x^3 + \cdots$
3. $f' = g$, $g' = -f$
4. From the previous question, $f'' = g' = -f$ so $f'' + f = 0$. Similarly $g'' + g = 0$.
5. The series converges for $-1 < x < 1$, by the ratio test, so it is differentiable for $-1 < x < 1$. We then have $f'(x) = 1 - x^2 + x^4 - x^6 + \cdots$, which is a geometric series with sum $1/(1+x^2)$.

Answers 6.6

1. $(f \circ g)(x) = (1+x)^2$, $(g \circ f)(x) = 1 + x^2$
2. $(f \circ g)(x) = \dfrac{1}{\sqrt{x+1} - 3}$, $(g \circ f)(x) = \sqrt{\dfrac{x-2}{x-3}}$
3. $(f \circ g)(x) = x + \sqrt{x}$, $(g \circ f)(x) = \sqrt{x^2 + x}$
4. $(f \circ g)(x) = (x+3)^2$, $(g \circ f)(x) = x^2 + 3$
5. $(f \circ g)(t) = \dfrac{2t^2 - 7}{t^2 - 4}$, $(g \circ f)(t) = \dfrac{1}{t}\left(4 + \dfrac{1}{t}\right)$
6. $(f \circ g)(x) = 125$, $(g \circ f)(x) = 5$
7. $4x(1+x^2)$
8. $12x^2(1+x^3)^3$
9. $4(3t^2 + 4t)(t^3 + 2t^2 + 1)^3$
10. $6(1 - 1/\phi^2)(\phi + 1/\phi)^5$

Answers 6.7

1. $-3e^{-3x}$
2. $-2xe^{-x^2}$
3. $\frac{1}{2}(e^t - e^{-t})$
4. $12(1 - e^{3x})^3$
5. $\dfrac{2e^x}{(e^x + 1)^2}$
6. $3x^2 e^{2x} + 2x^3 e^{2x}$
7. $2(2x^2 + x + 1)e^{x^2 + 2 + 1}$
8. $\dfrac{xe^x}{(1+x)^2}$
9. 1.221
10. 0.368

11. $1 + 5 + \dfrac{5^2}{2!} + \cdots + \dfrac{5^{27}}{27!}$

12. Here is one of many possibilities: $f(x) = \begin{cases} 1, & -\infty < x < 0 \\ -1, & 0 < x < \infty \end{cases}$

Answers 6.8

1. 36000   2. 4096   3. 1485 mg   4. $y = 4e^{3x}$ 7. 34.14°C   8. $\simeq 3.93\,\text{kg}$

Answers 7.2

2. $\omega = \sqrt{20}$.   The displacement after 2 s is $0.05\cos 2\sqrt{20}$ m.

Answers 7.3

5. $2\sin x \cos x$
6. $2\cos 2t \cos 3t - 3\sin 2t \sin 3t$
7. $2x \cos x - x^2 \sin x$
8. $2e^{2\phi}\cos\phi - e^{2\phi}\sin\phi$
9. $-\dfrac{1}{1-\cos x}$
10. $6\sin x \cos x (1 + \sin^2 x)^2$
11. $-\sin x - \sqrt{3}\cos x$
12. $f'(x) = \cos x,\ x \ne 0;\quad f'(0)$ does not exist.
13. $f(x) = \cos x,\, g(x) = \sin x$

Answers 7.6

6. $2x \sec^2 x^2$
7. $\sec^2 x \sin 2x + 2\tan x \cos 2x$
8. $-\operatorname{cosec} 4x \cot 4x$
9. $\operatorname{cosec} x^2 - 2x^2 \operatorname{cosec} x^2 \cot x^2$
10. $\left\{\dfrac{\sin x_n}{x_n}\right\} \to 1$ as $x_n \to 0$

Answers 8.1

1. L
2. NL (yy' term)
3. L
4. L
5. NL (y^2 term)
6. L
7. NL ((x'')^2, $x'x$, x^2 terms)
8. L
9. $y = Ae^{-2x} + Be^{-x}$
10. $y = Ae^{-5x} + B$
11. $y = (A + Bx)e^{-x}$
12. $y = (A + Bx)e^{-2x}$
13. $y = Ae^{-3x} + Be^{x}$
14. $y = Ae^{-(1+\sqrt{13})x/2} + Be^{-(1-\sqrt{13})x/2}$
15. $y = 5 - 2e^{-3x}$
16. $y = 2e^{x} - e^{2x}$
17. $y = (A + Bx)e^{4x}$
18. $y = Ae^{-4x} + Be^{x}$

19. $y = Ae^{-2x} + Be^x$
20. $y = (A + Bx)e^{-x/2}$
21. $y = C + (A + Bx)e^{-x/2}$
22. $y = (A + Bx)e^{3x} + Ce^{-x}$
23. $y = (A + Bx)e^{-2x/3}$
24. $y = Ae^{3x/2} + Be^{-2x}$
25. $y = (A + Bx)e^{2x} + Ce^{-2x}$
26. $y = (A + Bx)e^x + Ce^{-x}$
27. $y = A + Be^{-3x} + Ce^{3x}$
28. $y = (A + Bx + Cx^2)e^{-x/2}$
29. $y = Ae^{-t} + Be^{2t}$
30. $y = Ae^{-4t} + Bte^{-4t}$
31. $y = Ae^{(-1-\sqrt{5})t/2} + Be^{(-1+\sqrt{5})t/2}$
32. $x = Ae^{t/2} + Be^{-4t}$
33. $z = Ae^{(1+3\sqrt{5})t/2} + Be^{(1-3\sqrt{5})t/2}$
34. $x = Ae^{-5t/2} + Bte^{-5t/2}$
35. $y = Ae^{2t} + Be^{3t}$
36. $y = Ae^{3t} + Bte^{3t}$
37. $y = Ae^{-2t} + Be^{-3t}$
38. $y = Ae^{-2t/3} + Be^{t/2}$
39. $y = Ae^{t/2} + Bte^{t/2}$
40. $u = Ae^{-7t/3} + Be^{-t}$

Answers 8.2

1. 2
2. -3
3. 2
4. $\sqrt{5}$
5. $2i$
6. -5
7. $\sqrt{13}$
8. $\sqrt{5} + \sqrt{10}$
9. $5\sqrt{2}$
10. $5\sqrt{2}$
11. $-\sqrt{10}(3 + 4i)$
12. $-5 - 15i$
13. $9 - 8i$
14. 0
15. $-1 + 7i$
16. $-1/10$
17. $\frac{1}{10} + \frac{1}{10}i$
18. $\frac{1}{2} + \frac{1}{10}i$
19. $-\frac{1}{10} - \frac{3}{10}i$
20. $\frac{13}{25} - \frac{9}{25}i$
21. $\sqrt{10}/5$
22. $4 \operatorname{cis} \frac{11\pi}{6}$
23. $2 \operatorname{cis} \frac{\pi}{3}$
24. $4\sqrt{2} \operatorname{cis} \frac{\pi}{4}$
25. $2\sqrt{2} \operatorname{cis} \frac{3\pi}{4}$
26. $3 \operatorname{cis} \frac{3\pi}{2}$
27. $\sqrt{2} \operatorname{cis} \frac{5\pi}{4}$
28. $\frac{1}{2}(-1 \pm i\sqrt{15})$
29. $\frac{1}{3}(-1 \pm i\sqrt{2})$
30. $-1, \frac{1}{2} \pm i\frac{\sqrt{3}}{2}$
31. $1, -\frac{1}{2} \pm i\frac{\sqrt{3}}{2}$
32. $\pm 1, \pm i$
33. $\pm \frac{1}{2} \pm i\frac{\sqrt{3}}{2}$

Answers 8.3

1. $e^{-3\sqrt{3}}$

Answers 8.4

1. $y = e^{-2x}(A \cos 2x + B \sin 2x)$
2. $y = e^{-x/2}\left(A \cos \frac{\sqrt{3}x}{2} + B \sin \frac{\sqrt{3}x}{2}\right)$
3. $y = A \cos \omega x + B \sin \omega x$
4. $y = e^{2x}(A \cos 3x + B \sin 3x)$

5. $y = e^{-x}(A\cos\sqrt{3}x + B\sin\sqrt{3}x)$
6. $y = A\cos 2x + B\sin 2x$
7. $y = e^{-x}(A\cos 2x + B\sin 2x)$
8. $y = e^x(A\cos x + B\sin x) + Ce^{2x}$
9. $y = A\cos 3t + B\sin 3t$
10. $y = e^{2t}(A\cos t + B\sin t)$
11. $y = e^{-t/3}(A\cos\frac{\sqrt{2}}{3}t + B\sin\frac{\sqrt{2}}{3}t)$
12. $y = Ce^t + e^{-t/2}(A\cos\frac{\sqrt{3}}{2}t + B\sin\frac{\sqrt{3}}{2}t)$
13. $y = Ce^t + e^{-t}(A\cos t + B\sin t)$
14. $y = e^t(A\cos 6t + B\sin 6t)$
15. $y = 2\cos 4t - \frac{1}{2}\sin 4t)$
16. $y = e^{t/2}(\sin t/2 - \cos t/2)$
17. $y = -2\cos t$

Answers 8.5

2. $y = \frac{5}{7}\sin 14t$ cm, $t = \pi/14$ sec
3. $k = 194444$ N/m, (or $k = 19841$ kg/m), $\lambda = 23333$ N sec/m (or $\lambda = 2381$ kg sec/m).
4. $a < 4$
5. $m > 1$
6. $x = \cos\omega t$
7. $\omega = 2\pi \times$ frequency; $k = \omega^2 m = 320$ kg s^{-2}
8. $x = e^{-3t}(\cos\sqrt{7}t + (3/\sqrt{7})\sin\sqrt{7}t;\ \gamma = 3$
   $x = e^{-4t} + 4te^{-4t};\ \gamma = 4$
   $x = -\frac{1}{3}(e^{-8t} - 4e^{-2t});\ \gamma = 5$
9. $x = -\frac{1}{3}(e^{-8t} - 4e^{-4t});\ \omega = 4$
   $x = e^{-5t}(1 + 5t);\ \omega = 5$
   $x = e^{-5t}(\cos\sqrt{11}t + \frac{5\sin\sqrt{11}t}{\sqrt{11}});\ \omega = 6$
10. $x(t) = \frac{1}{10}\cos(7t\sqrt{10}/5) - \frac{1}{14\sqrt{10}}\sin(7t\sqrt{10}/5)$
    $\simeq 0.103\sin(4.427t + 1.793)$. After 1.369 sec
11. $\omega_2^2 : \omega_1^2$
12. Guitar string, pendulums, tides, alternating electric current, midday position of the sun as the seasons change.
13. Shock absorbers, resistance in an AC circuit.

Answers 8.6

1. $y = Ae^x + Be^{-2x} - \frac{3}{2}\cos x + \frac{1}{2}\sin x$
2. $y = A\cos x + B\sin x - \frac{1}{3}\sin 2x$
3. $y = Ae^{-x} + Be^x - \frac{\sin x}{2}$

4. $\cos 3x + e^{x/2}\left(A\cos\dfrac{\sqrt{35}x}{2} + B\sin\dfrac{\sqrt{35}x}{2}\right)$

5. $y = Ae^{-x}\cos 2x + Be^{-x}\sin 2x + \dfrac{3}{17}\sin 2x - \dfrac{12}{17}\cos 2x$

6. $x = A\cos t + B\sin t + \dfrac{1}{2}t\sin t$

7. $y = 2\cos x + 3\sin x - 2x\cos x$

8. $y = \dfrac{A}{2\omega^2}\sin\omega t - \dfrac{A}{2\omega}t\cos\omega t$

9. $y = \dfrac{A}{\omega^2 - \lambda^2}(\cos\lambda t - \cos\omega t)$

10. $x(t) = Ae^t + Be^{3t} + \dfrac{17}{3}$
11. $x(t) = A\cos t + B\sin t + 3e^t$

12. $x(t) = Ae^t + Be^{2t} + 3e^{3t}$
13. $x(t) = Ae^{-3t} + B + 2t$

14. $x(t) = Ae^{3t} + Be^{-2t} - \dfrac{1}{3}$
15. $x(t) = Ae^{-t} + Be^{-2t} - 5te^{-2t}$

16. (a) $a = 2$, (b) $x(t) = Ae^{-4t} + Be^{-t} + 2$, so $x(t) \to 2$ as $t \to \infty$. The transient part of the solution is $x(t) = Ae^{-4t} + Be^{-t}$ and the steady state part is $x(t) = 2$.

17. The initial value problem describing the motion is $30x''(t) + 750x = 20\sin 2t$, $x(0) = 0$, $x'(0) = 0$. The solution to this is

$$x(t) = -\dfrac{4}{315}\sin 5t + \dfrac{4}{63}\sin 2t.$$

18. Regular pushing on a swing. Unwanted feedback in public address systems. Tuning a radio: by tuning a dial, the natural frequency in the receiving circuit is made equal to the frequency of the broadcast waves of the sending station. The circuit resonates with the transmitted signals and absorbs peak energy from the signal.

19. (a) $q(t) = 2 - e^{-2t}\left(2\cos 3t + \dfrac{4}{3}\sin 3t\right)$; $i(t) = Ae^{-2t}\sin 3t$

    (b) $q(t) = \dfrac{1}{104}(13e^{-2t} - 8e^{-3t} - 5\cos 2t + \sin 2t)$; $i(t) = Ae^{-3t}(e^t - 1)$

20. $1000\,\Omega$

Answers 9.3

1. $x^4/4 + x^2 + x + c$
2. $e^{4x}/4 - \tfrac{1}{2}\cos 2x + c$

3. $\tan x - \tfrac{1}{3}\cos 3x + c$
4. $-\dfrac{1}{2x^2}$

5. $x^4/4 - x^3/3 + c$
6. $(x-2)^3/3 + c$

7. $t^3 + t + 3$

Answers 9.5

2. Yes

Answers 9.6

1. 155
2. 686
3. 506
4. 152
5. 259
6. 518
7. 44000
8. $\frac{1}{6}n(2n^2 - 3n + 1)$
9. $\frac{1}{6}p(2p^2 + 3p + 7)$
10. $\frac{1}{2}, \frac{1}{3}, \frac{1}{4}$
11. 28
12. 125
13. 124
14. 155
15. 9
16. 42

Answers 9.7

1. $T_n = 8.75$, $S_n = 8.6667$
2. $T_n = 0.4586$, $S_n = 0.4597$
3. $T_n = 0.7828$, $S_n = 0.7854$
4. $T_n = 1.2182$, $S_n = 1.21895$
5. $T_n = 2.4775$, $S_n = 2.4783$
6. $T_n = -155$, $S_n = -166.67$
7. The area of the cross-section is 11.1 m^2; the volume of water per second is 5.51 m^3/sec.
8. 2.6498 (6 subintervals)
9. 19 intervals
10. 20 intervals

Answers 9.8

1. $1 + x^2$
2. $1 + \sin 2x$
3. $x - 1/x^2$
4. $t + e^{3t}$
5. $2x(2x^2 + 1)$
6. $4u \sin 4u^4$
7. $2x(2x^2 + x^4)^2$
8. $3x^2 \alpha(x^3)$

Answers 9.9

1. $\dfrac{2}{2x + 4}$
2. $\dfrac{2x}{x^2 + 1}$
3. $\dfrac{4}{1 + x}$
4. $\dfrac{1}{x \ln x}$
5. $2x \ln(x^2 + 1) + \dfrac{2x^3}{x^2 + 1}$
6. $\dfrac{1 + 2x^2}{x(1 + x^2)}$
7. $\dfrac{2}{1 - x^2}$
8. $\dfrac{2x}{1 - x^4}$
9. $\dfrac{2x}{1 + x^2} + \dfrac{3x^2}{1 + x^3}$
10. $\dfrac{2(7x^2 + 2x - 3)}{(1 + 2x)(x^2 - 1)}$
11. $\dfrac{1 - e^{-x}}{x + e^{-x}}$
12. (a) $\simeq 2.5$. (In fact 2.512) (b) A straight line

13. (b) 1.16×10^{13}
14. (c) About 20 000 kilograms
15. (a) 10 Watts/m^2, (b) 130 Db

Answers 10.1

1. $f^{-1}(x) = \dfrac{x-3}{2}$
2. The function is not 1–1.
3. The function is 1–1 and so has an inverse. It is difficult to find a formula for the inverse. Newton's rule can be used to compute function values.
4. The function is not 1–1. This can be proved by showing that $x + 1/x = b$ has two solutions for $|b| > 2$.
5. $f^{-1}(x) = \left(\dfrac{1}{x} - 1\right)^{1/3}$
6. $f^{-1}(x) = \dfrac{3x+1}{x-1}$
7. f is not 1–1
8. $f^{-1}(x) = \dfrac{x-2}{5}$
9. $f^{-1}(x) = (x+1)^{1/3}$
10. $f^{-1}(x) = 1 - x^{1/3}$
11. $f^{-1}(x) = \dfrac{x}{1-x}$

Answers 10.2

1. x
2. x
3. x^2
4. $1/x$
5. e^x/x^2
6. $1 + \dfrac{1}{x}$
7. $x^x(\log x + 1)$
8. $x^{\sin x}(\cos \log x + \dfrac{\sin x}{x})$
9. $(\cos x)^x (\log \cos x - x \tan x)$
10. $\dfrac{2^{\sqrt{x}} \log 2}{2\sqrt{x}}$
11. $\dfrac{(\log x)^{\sqrt{x}}}{2\sqrt{x}} (\log \log x + \dfrac{2}{\log x})$
12. 1.41421
13. 1.25992
14. 0.739085
15. 1.86081
16. 0.876726
17. 3.44949
18. 1.47502
19. 1.25992
20. 0.346574
21. 20.0855
22. 0.523599
23. 2.44949
25. $\dfrac{10 \log 2}{\log 6} \simeq 3.9$hrs
26. Take $k = \dfrac{3.8}{5} \times 10^{-4}$day^{-1}. The population doubles after 25 years and triples after 39.6 years.
27. 3.817 days, 8.86 days
28. 37 000 years
29. About 14 minutes
30. 10.08 pm

Answers 10.5

1. $\pi/2$
2. $\pi/2$
3. $\pi/3$
4. $\pi/3$
5. $-\pi/4$
6. $\sqrt{3}/2$
7. $\sqrt{2}/2$
8. $\pi/4$
9. $\pi/2$
10. $-\pi/4$
11. 0
12. $1/\sqrt{2}$
13. $\dfrac{x}{\sqrt{1+x^2}}$
14. $\dfrac{1}{\sqrt{1+x^2}}$
15. x
16. $\sqrt{1-x^2}$
17. x
18. $\dfrac{\sqrt{1-x^2}}{x}$
19. x
20. x
21. x
22. $f'(x) = \dfrac{-2x}{\sqrt{1-(x^2-1)^2}}$
23. $f'(x) = \dfrac{-1}{x^2+1}$
24. $f'(x) = e^x \left(\cos^{-1} x - \dfrac{1}{\sqrt{1+x^2}} \right)$
25. $\dfrac{dy}{dx} = \dfrac{\cos x}{1+\sin^2 x}$
26. $\dfrac{dy}{dx} = \dfrac{x^2}{(4-x^2)^{3/2}}$
27. $\dfrac{dy}{dx} = \dfrac{3-x}{\sqrt{9-x^2}}$
28. $f'(r) = \dfrac{r}{|r|\sqrt{1-r^2}}$
29. $\dfrac{2x^{2\,x}(1+\log x)}{1+x^{4\,x}}$

Answers 11.1

1. $y(t) = A\cosh\sqrt{2}t + B\sinh\sqrt{2}t$
2. $y(t) = A\cosh 3t + B\sinh 3t$
3. $y(t) = e^{-3t}(A\cosh 2t + B\sinh 2t)$
4. $x(t) = e^t(A\cosh\sqrt{3}t + B\sinh\sqrt{3}t)$

Answers 11.2

13. $f'(x) = 2\sinh(2x+1)$
14. $f'(x) = 3x^2 \cosh x^3$
15. $f'(x) = \operatorname{csch} 2x$
16. $f'(x) = 2\coth 2x$
17. $f'(x) = 3\cosh^2 x \sinh x$
18. $f'(x) = \tanh^2 x$
19. $f'(x) = 3x^2 \sinh x^3$
20. $f'(x) = \dfrac{1}{x}\sinh(\log x)$
21. $f'(x) = -\tanh x$
22. $f'(x) = -\dfrac{1}{x^2}\cosh\dfrac{1}{x}$
23. $f'(x) = \cos x \cosh(\sin x)$
24. $f'(x) = \dfrac{1}{1-\cosh x}$
25. $f'(x) = a\coth ax$
26. $f'(x) = \operatorname{sech}^2\left(\dfrac{x}{2}\right)$

27. (b) The limiting velocity is $\sqrt{mg/k}$. This is the *terminal velocity* of the body. No matter how far it falls, it never exceeds this velocity.

(c) $s(t) = \dfrac{m}{k} \log \cosh \left(\sqrt{\dfrac{kg}{m}} t \right) + s_0$

Answers 11.3

1. $f'(x) = \dfrac{2}{\sqrt{4x^2 - 1}}$
2. $f'(x) = \dfrac{2x}{\sqrt{1 + (1 + x^2)^2}}$
3. $f'(x) = -\dfrac{1}{(1 - x^2) \tanh^{-1} x}$
4. $f'(x) = \dfrac{1}{\sqrt{x^2 + 1}} + \dfrac{1}{\sqrt{x^2 - 1}}$
5. $f'(x) = \dfrac{a \sinh x}{\sqrt{a^2 \cosh^2 x + 1}}$
6. $f'(x) = \dfrac{3(\sinh^{-1} x)^2}{\sqrt{1 + x^2}}$
7. $f'(x) = \dfrac{1}{(1 - x^2) \tanh^{-1} x}$
8. $f'(x) = \operatorname{sech} x$
9. $f'(x) = \dfrac{x}{|x|\sqrt{x^2 - 1}}$
10. $f'(x) = \dfrac{x}{|x|\sqrt{x^2 + 1}}$

Answers 12.2

1. 1
2. $\sqrt{2} - \pi/4$
3. 93
4. $\pi/4$
5. $\pi/2$
6. $\sinh 4$
7. 0
8. $\sqrt{3} - 1$
9. 21/64
10. 2
11. 4
12. 2

Answers 12.3

Note: The constant of integration has been omitted in the answers below.

1. $\dfrac{(x^2 + 1)^2}{2}$
2. $-\dfrac{\cos^4 x}{4}$
3. e^{x^2}
4. $\sqrt{x^2 + 1}$
5. $\dfrac{2}{3}(x + 1)^{3/2}$
6. $\dfrac{2}{9}(x^3 + 4)^{3/2}$
7. $\dfrac{\sin x^3}{3}$
8. $-2 \cos \sqrt{x}$
9. $\dfrac{2\sqrt{a + bt^n}}{bn}$
10. $\dfrac{1}{3}(as^2 + 2bs + c)^{3/2}$
11. $\dfrac{(\log x)^2}{2}$
12. $-e^{\cos \theta}$
13. $\dfrac{1}{2} \log(1 + \sinh 2x)$
14. $\log |\sinh u|$
15. $-\dfrac{u\sqrt{1 - u^2}}{2} + \dfrac{\sin^{-1} u}{2}$
16. $\dfrac{3}{2}(\log t)^2$
17. $\dfrac{1}{2} \log(\log x^2)$
18. $\dfrac{(\tan^{-1} x)^2}{2}$
19. $-\tanh^{-1} \sqrt{1 - e^{2p}}$
20. $\sin^{-1}(\sin x)$

Answers 12.4

Note: The constant of integration has been omitted in the answers below.

1. $xe^x - e^x$
2. $-x\cos x + \sin x$
3. $(2 - x^2)\cos x + 2x\sin x$
4. $x\sin^{-1} x + \sqrt{1 - x^2}$
5. $x\tan^{-1} x - \dfrac{1}{2}\log(1 + x^2)$
6. $\dfrac{x^2}{3}\sin^{-1} x + \dfrac{1}{9}(2 + x^2)\sqrt{1 - x^2}$
7. $-\dfrac{e^x}{5}(2\cos 2x - \sin 2x)$
8. $-\dfrac{x^3}{9} + \dfrac{x^3}{3}\log x$
9. $-\dfrac{\sqrt{4+x}}{x} - \dfrac{1}{2}\tanh^{-1}\left(\dfrac{\sqrt{4+x}}{2}\right)$
10. $\log(\cos x) + x\tan x$
11. $-2x + 2\sqrt{1 - x^2}\sin^{-1} x + x(\sin^{-1} x)^2$
12. $-\dfrac{1}{x} - \dfrac{\log x}{x}$
13. $-\cos(\log(\sin x)) + \log|\cosec x - \cot x| + \cos x$
14. $\dfrac{2x^3}{27} - \dfrac{2x^3}{9}\log x + \dfrac{x^3}{3}(\log x)^2$

Answers 12.5

Note: The constant of integration has been omitted in the answers below.

1. $\log\left(\dfrac{1+x}{2+x}\right)$
2. $\dfrac{1}{2}\log\left(\dfrac{1+x}{x-1}\right)$
3. $\log\left(\dfrac{x-2}{2x+1}\right)$
4. $\dfrac{1}{5}\log\left(\dfrac{x-3}{x+2}\right)$
5. $\dfrac{\log x}{2} - 2\log(1 + x) + \dfrac{5}{2}\log(2 + x)$
6. $\dfrac{1}{2}\tan^{-1} x + \dfrac{1}{4}\log\left(\dfrac{x-1}{x+1}\right)$
7. $\log x - \dfrac{1}{2}\log(1 + x^2)$
8. $\tan^{-1} x + \log\left(\dfrac{1+x}{\sqrt{1+x^2}}\right)$
9. $\dfrac{\sqrt{2}}{3}\tan^{-1}\dfrac{x}{\sqrt{2}} + \dfrac{1}{6}\log\left(\dfrac{x-1}{x+1}\right)$
10. $\dfrac{7}{3}\log\left(\dfrac{x-5}{x-2}\right)$
11. $-\dfrac{1}{2}\log(1 + x) + 2\log(2 + x) - \dfrac{3}{2}\log(3 + x)$
12. $-\dfrac{1}{\sqrt{3}}\tan^{-1}\left(\dfrac{1+2x}{\sqrt{3}}\right) + \log x - \dfrac{1}{2}\log(x^2 + x + 1)$
13. $-\dfrac{\pi}{4} + \tan^{-1} 2$
14. $\log\left(\dfrac{1+\cos x}{2+\cos x}\right)$
15. $6t^{1/3} + 3\log\left(\dfrac{t^{1/3} - 1}{t^{1/3} + 1}\right) + t$
16. $5\log\left(\dfrac{u^{1/5}}{1+u^{1/5}}\right)$

Answers 12.7

Note: The constant of integration has been omitted in the answers below.

1. $\dfrac{1}{2} \tan^{-1}\left(\dfrac{4+x}{2}\right)$

2. $-\dfrac{1}{\sqrt{3}} \tan^{-1}\left(\dfrac{1+2x}{\sqrt{3}}\right) + \dfrac{1}{2} \log(x^2 + x + 1)$

3. $\dfrac{1}{8} \log(2x - 1) + \dfrac{3}{8} \log(2x + 7)$

4. $\dfrac{2}{\sqrt{3}} \tan^{-1}\left(\dfrac{1+2e^x}{\sqrt{3}}\right)$

5. $\dfrac{1}{2} \tan^{-1}\left(\dfrac{\sin x - 1}{2}\right)$

6. $\dfrac{1}{18}\left(7 \tan^{-1}\left(\dfrac{1+3x}{2}\right) + 2\log(9x^2 + 6x + 5)\right)$

7. $\tan^{-1}(2+x) + \dfrac{1}{2} \log(x^2 + 4x + 5)$

8. $-\dfrac{1}{\sqrt{3}} \tan^{-1}\left(\dfrac{1+2x}{\sqrt{3}}\right) + \dfrac{3}{2} \log(x^2 + x + 1)$

9. $-\dfrac{2}{\sqrt{3}} \tan^{-1}\left(\dfrac{1+2x}{\sqrt{3}}\right) + \log x$

10. $\tan^{-1}(2 + \tan x)$

Index

Bold face page numbers are used to indicate pages with the primary information about the entry, such as a definition or an explanation.

abuse of notation, 58
acceleration, 55
alternating series, 114
amplitude
    of a complex number, 138
    of SHM, 146
antiderivatives, 169
Anzac Bridge, 7, 9, 109
Archimedes, 176
area under a curve, 171
Argand, 136
Argand diagram, 135
argument
    of a complex number, 138
Aristotle, 4
asymptotes, 23
atmospheric pressure, 6
auxiliary equation, 129
    complex roots, 133
    distinct roots, 130
    repeated roots, 132

backward differences, 61
barometer, 5
beats, 159

Cardano, 134
central differences, 61, 76
chain rule, 96
complementary
    equation, 153
    function, 153
complex
    conjugate, 138
    exponential, 141

    plane, 135
    series, **140**
complex number, **134**
    amplitude, 138
    argument, 138
    Cartesian form, 138
    imaginary part, 135
    modulus, 138
    phase, 138
    polar form, 138
    real part, 135
composition of functions, 96
concave down, 72
concave up, 72
concavity, 73
conjugate pairs, 139
continuity, 28
    at a point, 33
    on a closed interval, 37
    on an open interval, 37
continuous, 28
continuous function, **33**, 37
convergent sequence, 32
convergent series, 82
cosecant, 125
cosine function, 113
cotangent, 125
critical damping, 149

damping, 145
de Moivre's theorem, 143
decreasing function, 68, 69
derivative, **45**
    of $\log x$, 188
    of $\cos x$, 116
    of $\sin x$, 116
    of $\tan x$, 125
    of x^n, 52, 204
    of a constant function, 49

of a product, 50
of a quotient, 52
of a square root, 204
of a sum, 49
of an inverse function, 204
of inverse trigonometric functions, 219
differences
    backward, 61
    central, 61, 76
    forward, 61
differentiable, 41, 44
differential equation, 61
    homogeneous linear, 129
    linear, 129
    nonlinear, 129
    second order, 111
    solved using *Mathematica*, 67
differentiating a power series, 93
differentiation
    by first principles, 45
    logarithmic, 193
discontinuous, 28
domain, 12
domain, natural, 12, 13
Dreamworld, 3

elementary function, 225
elementary functions, 252
elliptic functions, 254
energy, 215
error term, 114
Euler, 141
Euler's formula, 141
exponential function, 87
    general, 212
extreme value, 70

finite differences, 46
flight 811, 4
fluid pressure, 5
forced oscillations, 156
forward differences, 61
frictional forces, 128
function, 8, **12**
    continuous, 32, **33**, 37
    decreasing, 69
    differentiable, 28, **44**
    graph of, 17

    increasing, 69
    primitive, 169
    rational, 25, 57
    smooth, 27, 28
    value, 12

Galileo, 58, 66
general exponential function, 212
general solution, 103, 130
geometric series, 82
graph
    of a function, 17
    plotting with *Mathematica*, 17

hectopascal, 75
homogeneous, 129
hyperbolic functions, 225, 227

increasing function, 68, 69
indefinite integral, 170
independent variable, 12
infinite series, 81
initial value problem, 63
input number, 11, 12
integral, 170
    indefinite, 170
integration by parts, 241
integration by substitution, 239
intersection, 15
interval
    closed, 15
    infinite, 15
    open, 15
intervals, 15
inverse function, 197, 200, **201**
inverse function values, calculating, 205
inverse hyperbolic functions, 230
inverse sine, **218**
inverse trigonometric functions, **218**

Jacobi amplitude, 256

kinetic energy, 215

left-hand sum, 177
Leibniz, 45
linear
    combination, 130

differential equation, 128
local maximum, 70
local minimum, 70
logarithmic differentiation, 193

maximum
    absolute, 71
    global, 71
    local, 70
method of undetermined coefficients, 155
millibars, 75
minimum
    absolute, 71
    global, 71
    local, 70
modulus
    of a complex number, 138

natural domain of a function, 13
natural domain of a rule, 12
Newton, 45
Newton's method, **206**
Newton–Raphson method, 206
nonlinear, 252
numerical properties, 8

Olympic Games, 1
one–to–one, 200
output number, 11, 12

partial fractions, 243
partial sums, 82
particular integral, 153
particular solution, 103, 153
parts, integration by, 241
pascal, 75
pendulum, simple, 251
periodic, 122
periodic functions, 116
periodicity, 116
phase
    of a complex number, 138
phase angle, 146
physical laws, 66
point of inflexion, 71, 73
    horizontal, 71
polar form
    of a complex number, 138

polynomial, 21
potential energy, 215
primitive, 170
primitive function, 169
principle of superposition, 130
products of functions, 50
projectile motion, 58
pump, 7

quadratic integrals, 247
quotients of functions, 50

radius of convergence, 90
range, 12
Raphson, 206
rate of change, 54
ratio test, 85, 141
rational function, **25**
rational functions, 57, 244
recurrence relation, 112
resonance, 158
restriction, 12
Riemann sum, 176
right-hand sum, 177

secant, 125
second derivative, 45
sequence, **32**
    convergent, 32
    divergent, 32
simple harmonic motion, 145, 146
simple pendulum, 251
Simpson, 206
Simpson's rule, 183
sine function, 113
solution
    general, 103, 130
    of a differential equation, 61
    particular, 103
    trial, 62
square wave, 30
stationary point, 71
steady state, 161
stretched string, 127
strong damping, 148
substitution, integration by, 239
sum
    Riemann, 176

sums of functions, 49
superposition, principle of, 130

Tacoma Narrows Bridge, 7
tangent, 125
Torricelli, 5
Tower of Terror, 1, 57
transient, 160
trapezoidal rule, 182
trial solution, 62, 104
trigonometric functions as ratios, 119
trigonometric identities, 116
turning point, 71

undetermined coefficients, 154
union, 15

variable, 9
variation of parameters, 104
velocity, 53

weak damping, 147